D0825219

Remarkable Biologists

Following on from the success of his t
Mathematicians and *Remarkable Physicists,* Ioan James now profiles
thirty-eight remarkable biologists from the last 400 years. The emphasis is
on their varied life stories, not on the details of their achievements, but
when read in sequence their biographies, which are organised
chronologically, convey in human terms something of the way in which
biology has developed over the years. Scientific and biological detail is
kept to a minimum, inviting any reader interested in biology to follow this
easy path through the subject's modern development.

IOAN JAMES is Emeritus Professor of Mathematics at the University of
Oxford and has had a distinguished career as a research mathematician. In
recent years he has become interested in the history and development of
scientific disciplines and the scientists involved. He was elected Fellow of
the Royal Society in 1968.

Remarkable Biologists

Ioan James

University of Oxford

CAMBRIDGE
UNIVERSITY PRESS

CAMBRIDGE UNIVERSITY PRESS
Cambridge, New York, Melbourne, Madrid, Cape Town, Singapore, São Paulo, Delhi

Cambridge University Press
The Edinburgh Building, Cambridge CB2 8RU, UK

Published in the United States of America by Cambridge University Press, New York

www.cambridge.org
Information on this title: www.cambridge.org/9780521875899

First published 2009

Printed in the United Kingdom at the University Press, Cambridge

A catalogue record for this publication is available from the British Library

Library of Congress Cataloguing in Publication data
James, I. M. (Ioan Mackenzie), 1928 –
Remarkable biologists / Ioan James.
p. cm.
Includes bibliographical references.
ISBN 978-0-521-87589-9 (hardback)
1. Biologists – Biography. 2. Biology – History. I. Title.
QH26.J36 2009
570.92′2 – dc22
[B] 2008053620

ISBN 978-0-521-87589-9 hardback
ISBN 978-0-521-69918-1 paperback

Contents

Preface

This book is intended for those who would like to read something, but not too much, about the life stories of some of the most remarkable biologists born in the last four hundred years. Biology is a many-sided subject and I interpret the term biologist quite loosely. I begin in the seventeenth century, and end in the twenty-first, but exclude living people as a matter of principle. There are four or five subjects in each of nine chapters, making thirty-eight profiles altogether. The emphasis is mainly on their varied life stories, less on the details of their achievements. The mini-biographies are arranged chronologically by date of birth, so that when read in sequence they convey in human terms something of the way in which the discipline developed. It has not been easy to select a limited number of individuals from far more possibilities, but I mainly choose those for whom a full-scale biography exists. The emphasis is on variety, certainly not on making a list of the most important biologists of the past four centuries. The subjects I have chosen come from ten different countries. Some of them advanced the subject through their research or scholarship, others found some aspect of biology fascinating and studied it from sheer pleasure.

Prologue

Biology is the study of living organisms. More than any other branch of science it has enjoyed a broad appeal, throughout the ages. We all want to know something about the plants and animals which inhabit the natural world, or have done so in the past. We try and identify the birds and other creatures we see around us. We find that as well as the macrocosm of creatures visible to the naked eye there is a microcosm of bacteria, protozoa and other tiny organisms, only visible under the microscope, and that some of these cause disease. All this is common knowledge but specialists in the field, and a host of enthusiastic amateurs, know far more. The various branches of biological science complicate any account of the development of the subject. Nordenskiold (1929) has given an overview but each branch has its own distinctive history and its own culture.

Biology grew out of natural history (which is not history in the modern sense of the word: natural science is a more satisfactory alternative). Pliny the Elder (ca. 22–78) compiled a comprehensive and well-organised guide to the huge amount of information about the wonders of nature that was available in antiquity. He called this *Historia naturalis*, hence the term natural history; practitioners are called *naturalists*. Pliny drew on the work of his Greek predecessors, such as Aristotle and his disciple Theophrastus. Many copies of the *Historia naturalis* were made during the Middle Ages; the first printed edition appeared in 1469, and many others appeared later. Translations were made of the three zoological books of Aristotle and the two botanical books of Theophrastus, but scholars began to believe that it would be better to study animals and plants directly than by reference to these ancient sources. One of the last, and perhaps the greatest, of these encyclopaedic naturalists in the tradition of Pliny was Conrad Gessner (1516–1565), of Zurich, who died of the plague before his projected *Historia plantarum* was written, but whose *Historia animalium*, which appeared in 1551–1558, is regarded as the beginning of modern zoology.

Pliny's near contemporary Dioscorides (ca. 40–80) wrote about the importance of understanding the natural world in light of its medicinal efficacy, describing approximately 550 Mediterranean plants which had uses in medicine. In the sixteenth century natural history began to form part of the medical curriculum at universities, which often maintained a botanical or physic garden where medical students could be shown plants with medicinal uses. The herbarium (a collection of dried plants) was also used to demonstrate the different kinds of plants. Professors of medicinal botany sometimes took the students on field trips in the summer months. Another development was the formation of cabinets of curiosities, some of which developed into museums. There was a market for books on natural history, often with illustrations, but only a few of the artists who provided these were naturalists themselves. In the seventeenth century the invention of the microscope opened up a new and completely unexpected world. That the smallest drop of water contained countless living creatures was beyond the imagination. It also made it possible to examine the detailed structure of parts of the human body.

The seventeenth century was also the age of the early voyages of exploration. Travellers from England, France, Holland, Portugal, Russia and Spain began to explore the more accessible parts of the world, sometimes by land but usually by sea. Their main purpose was commercial and strategic, rather than scientific. The discovery of the New World at the end of the fifteenth century was followed by a stream of plants and animals unlike those which were already familiar to Europeans. Although few naturalists crossed the Atlantic themselves specimens were collected on their behalf and brought to Europe. Clusius, for one, formed a famous collection of such material at the University of Leiden. As further voyages of exploration were sent out, it became more usual for a naturalist to be taken along, to make observations and to collect interesting material; someone who was medically qualified might fulfil this role. On scientific voyages there was usually a draughtsman on board, whose work might be used to illustrate the published account of the voyage, if one was prepared. Once regular shipping routes were established, as they were to Latin America, it became easier for naturalists to visit the places from which they had been receiving specimens, if they could afford to do so. Those who wished to travel could usually defray their expenses by collecting and importing material which could be sold to naturalists at home.

The great voyages of the latter part of the seventeenth century were usually primarily for a specific objective, for example to search for a northwest passage into the Pacific Ocean or to find the *terra australis incognita*, the continent that since classical times had been thought necessary to balance the land mass in the northern hemisphere. Scientific programmes were usually secondary to the strategic concerns of governments in Britain, France, Russia and Spain. There were political problems which made international collaboration often hard to achieve. The Dutch made it difficult to reach the Pacific via the Cape of Good Hope, while Spain guarded entry around Cape Horn via the Straits of Magellan.

For a number of reasons voyages of discovery might be connected to what might loosely be called scientific expeditions, pioneering examples of which were sponsored by the Académie Royale des Sciences in the eighteenth century, notably the expeditions to Lapland and Peru to settle the question as to whether or not the Earth was flattened at the poles. One of the most celebrated voyages was that of Louis Antoine de Bougainville, who had founded a settlement on the Falkland Islands on an earlier voyage in 1764. He returned two years later, accompanied by the able naturalist Philibert Commerson (1727–1773), reached Tahiti through the Straits of Magellan, and went on to Samoa, the New Hebrides, the Moluccas, Java and Mauritius. From there the expedition returned to France, after having been away three years, while Commerson was left to explore Mauritius and Madagascar before returning ill to Mauritius, where he died. Commerson collected over 3000 new species which were added to the collections of the Jardin du Roi in Paris; after traversing most of the Pacific rim the expedition foundered on the reef of Vanikoro, off the Santa Cruz Islands, in early 1788. At about the same time the Royal Society of London sought information from voyagers about ethnography and natural history on voyages sponsored by the Royal Navy. Cook's voyages prompted French and Spanish competition, which in turn prompted the Admiralty to send out further expeditions with frankly strategic objectives. The Spaniards invested a great deal of money in botanical expeditions over the last few decades of the century but refused to publish what was discovered, and the Russians also were secretive.

During the nineteenth century the discoveries of science increasingly came into conflict with religious beliefs. Geologists had long believed that the Earth must be far older than stated in the Bible. Fossils were collected and examined. Some appeared to be the remains of organisms which no longer existed; what did this mean? How did it come about that fossils of marine organisms could be found at high altitudes? Instead of there being one act of creation, the idea that there might have been a series of separate acts of creation gained ground. One theory was that the history of the Earth had been marked by great catastrophes, of a scope and intensity never witnessed by man, which wiped out large portions of the animal world. These were then replaced by other forms of life in another act of creation. Those who held this view were known as catastrophists, as distinct from the uniformitarians, who maintained that the processes at work in historical times, which produced such features as mountain ranges, volcanoes and glaciers, could be taken as a guide to what might have happened long ago.

There were also different theories as to how the natural world came about. The almost unlimited time required by the geologists for the development of the surface of the Earth made the biologists think again, and various evolutionary theories were proposed. Although the basic idea of evolution was widely accepted most scientists believed, well into the nineteenth century, that the natural world had been made by a purposeful Creator in successive independent creations. Although the general principle of the evolution of species was accepted by most biologists, there was a tendency to follow Lamarck and believe in the inheritance of acquired characteristics. This was the situation when Darwin, and independently Wallace, presented the case for natural selection being the main driving force for evolution. Both Darwin and Wallace were unaware that the Austrian monk Gregor Mendel was elucidating the laws of genetics which provided an essential ingredient of the theory of natural selection. For a convincing theory of evolution some explanation of the mechanism was required. Genetics provided this and the idea emerged that genes are the keys to the situation. Today there are still clashes between different schools of thought in this area.

Most branches of science experienced exponential growth in the twentieth century and biology is no exception. New branches of the discipline developed, of which molecular biology is an important example. The determination of the structure of DNA by Crick and Watson led to a better understanding of some of the fundamental problems of biology but others remain. Evolutionary biology still provokes controversy, as does eugenics. Conservation is another area where biologists play a leading role. In the final chapters I have included some comparatively short profiles to illustrate at least some aspects of this variety of activity.

1 From Ray to Leeuwenhoek

JOHN RAY (1627–1705)

I begin in seventeenth-century England with a naturalist whose writings earned him the titles of 'the father of natural history' and 'the Aristotle of England and Linnaeus of the time'. Cuvier said that his works were the basis of all modern geology, von Haller that he was the greatest botanist in the memory of man. John Ray, the third child of Roger Ray, the village blacksmith, and his wife Elizabeth was born in the hamlet of Black Notley in rural Essex on November 29, 1627. Elizabeth Ray 'was of great use in her neighbourhood,' we are told, 'particularly to her neighbours that were lame or sick, among whom she did a great deal of good, especially in chirurgical matters'. Her son John was educated at Braintree School, where his ability was recognised, and at the age of sixteen went to Cambridge, supported by a bursary from a local squire. After two years at the Catharine Hall he migrated to Trinity College, where there was less scholasticism. The brilliant Isaac Barrow was a fellow-pupil; their tutor described them as more able than Isaac Newton was as an undergraduate. After graduation Ray was elected to a minor fellowship and began the study of botany. In the following decade he held a series of college offices. Ray was a prolific writer all his life; his first book, published in 1660, was a flora of the neighbourhood of Cambridge. D'Arcy Thompson, writing in 1922, says Ray describes localities so minutely that Cambridge students still gather some of their rare plants in the copses and chalk-pits where he found them centuries before.

Ray took holy orders at the end of 1660, but had to resign his fellowship two years later because he refused to subscribe to the 1661 Act of Conformity, the only fellow of Trinity to refuse and one of only twelve in the whole university. This left him without any professional means of livelihood. Ray had made a number of friends who shared his rejection of scholasticism in favour of a science based on observation and experiment. Some of them were elected to the Royal Society in the early years of its existence. One was Hans Sloane, whose profile follows later in this chapter, another was Robert Boyle. However the most important in Ray's life

was Francis Willughby (1635–1672), a gifted youth of great promise who died young. He was the only son of the baronet Sir Francis Willughby of Middleton Hall, Warwickshire, and of his wife, Lady Cassandra. Eight years younger than Ray he came up to Cambridge in 1652 and proved to be an exceptional student, particularly in mathematics. He also became interested in natural history, especially zoology. Later Willughby spent some time in Oxford studying at the Bodleian Library, and became one of the original fellows of the Royal Society at the age of twenty-five. He and Ray became close friends and he was able to provide Ray with much-needed financial support.

Ray's botanical expeditions started in 1659. At first he was on his own, travelling usually on horseback, sometimes on foot. Although botany was his primary purpose, he also liked to collect information about the places he visited. Later Willughby started to come with him; then his former pupil Philip Skippon joined them. Willughby proposed that the three of them should cross the English Channel and tour on the continent. They set off for France in 1663, travelling in stages through the Low Countries, then up the Rhine and down the Danube to Vienna. Then they crossed the Alps into Italy, where they spent some time in Padua studying anatomy. After visiting some of the other Italian cities they split up at Naples, with Willughby returning home via Spain while Ray and Skippon went on to Sicily and Malta before taking a more circuitous route which took them via Rome,

Venice and Geneva to Montpellier, then enjoying a golden age as a centre of intellectual life, especially medical and botanical studies. They reached home in the spring of 1666, the year of the Great Plague, having been away three years. On the tour Ray had visited some of the great continental centres of learning, met some of the leading experts in his field, and made himself familiar with the flora and fauna of Western Europe.

Willughby had paid Ray's expenses on their tours and now he also provided him with a base for his work at Middleton Hall, the residence of his parents. From there Ray brought out his *Catalogus Plantarum Angliae et Insularum Adjacentium* and *Collection of English Proverbs* and prepared several more books, including his *Observations Topographical, Moral and Philosophical*, containing a list of foreign plants, and his *Collection of English Words*, which contained much information about local dialects. He was elected to the Royal Society in 1667, and occasionally attended its meetings.

Willughby married the heiress Emma Barnard in 1668, and children followed rapidly, but he died four years later. In his will his benefactor and friend left Ray an annuity of £60, which freed him from financial worries, and entrusted Ray with the education of his sons. He left many papers on natural history, although he had published nothing. Ray felt it his duty to go through them and publish what he could, in Willughby's name, so for the next three and a half years he was occupied in preparing *Ornithology* and *Historia piscium* for publication. Ray also acted as resident chaplain for the household and tutor for the Willughby sons. They had a governess named Margaret Oakley who married Ray in 1673.

During the first six years of marriage they had no fixed home. Unfortunately Willughby's widow had taken a dislike to Ray and when Lady Cassandra died in 1675 he had to leave Middleton Hall. He needed to complete his work on Willughby's papers and so they found a nearby place to live. In 1677 he declined an invitation to become secretary of the Royal Society. The same year they moved to another house a few miles from Black Notley. When his father had died in 1656 Ray had provided for his 'most dear and honoured' widowed mother, to whom he was devoted, by having a house built for her at Black Notley, called Dewlands. When she died in 1679 he and Margaret moved into Dewlands and started a family; they had four children, all daughters, who helped him by collecting Lepidoptera and other material. Working alone in his library of 1500 volumes he produced a stream of botanical books, also several widely read religious works. One, *The Wisdom of God, Manifested in the Works of the Creation*, many times reprinted, was the source of William Paley's famous *Natural Theology*, which became a standard work in the nineteenth century. Another, *Persuasives to a Holy Life*, was based on sermons he had given in Cambridge long before.

Ray died on January 17, 1705, having been in poor health for some time. Throughout his adult life he corresponded regularly with his friends and with other naturalists. He was a man of great industry and had a flair for collecting information and arranging it systematically. According to Cuvier 'the particular distinction of his labours lies in an arrangement more clear and determinate than those of any of his predecessors and applied with more consistency and precision'. His great contribution to botany was the system of classification that he introduced, which although later superseded, produced order out of unsystematic descriptions. He applied similar principles to the very difficult classification of insects and other invertebrates. His last works, *Historia Insectorum* and *Synopsis Avium et Piscium*, were published posthumously.

MARIA SIBYLLA MERIAN (1647–1717)

There were plenty of women naturalists but it is hard to find out much about them. An exception is Maria Sibylla Merian, who is surely one of the earliest. She was born in the free imperial city of Frankfurt-am-Main on April 12, 1647, the first child of the artist and publisher Mathias Merian the Elder by his second wife Johanna Sibylla Heim. Mathias was then in his fifties, known throughout Europe for his engravings of cityscapes and landscapes, his scientific books, and his editions of the illustrated series *Grands voyages* (accounts of journeys to the New World) begun by his first father-in-law, Theodore de Bry. Mathias died when Maria Sibylla was only three, and her mother soon remarried. Her second husband was the widower Jacob Marrel, a still-life painter, engraver and art dealer.

Mathias Merian the Elder and Jacob Marrel had both acquired citizenship in Frankfurt, and Maria Sibylla could later claim her own Burgerrecht as Mathias' daughter. Both men enjoyed affluence and prestige and ranked as artists well above other craftsmen in Frankfurt's ordering of estates. They were immigrants, as was Maria Sibylla's mother; Mathias was a native of Basel; Johanna Sibylla came from a Walloon family that had migrated from the Netherlands to nearby Hanau; Marrel had a French grandfather who had moved to Frankfurt, but he himself had been born in the Palatinate town of Frankenthal and had spent years in Utrecht before settling in Frankfurt.

Maria Sibylla's half-brothers, Mathias the Younger and Caspar Merian, were establishing themselves as engravers, publishers and painters, producing topographic works in the tradition of their father, recording

MARIA SIBILLA MERIAN
Nat. XII. Apr. MDCXLVII. Obiit XIII. Jan. MDCCXVII.

ceremonial events like the coronation of the emperor Leopold I at Frankfurt, and much else. Almost all the women artists of the early period were, like Maria Sibylla Merian, born into families of artists. In that setting their talent could be welcomed, and contemporary beliefs about the dampening effects of the female temperament on genius could be disregarded. While her mother taught her embroidery, she was able to learn drawing, watercolour, still-life painting and copperplate engraving from her stepfather along with his male pupils. She was also able to study the large collection of prints, books and paintings belonging to Jacob Marrel and the Merrians.

Maria Sibylla's fascination with natural history began early: 'from my youth onward I have been concerned with the study of insects. I began with silkworms in my native city, Frankfurt am Main; then I observed the far more beautiful butterflies and moths that developed from other kinds of caterpillars. This led me to collect all the caterpillars I could find to study their metamorphoses . . . and to work at my painter's art so that I could sketch them from life and represent them in lifelike colours.' In 1665 she married Johann Andreas Graff of Nuremberg, a favourite pupil of her father's. After five years in Frankfurt, where their first daughter, Johanna Helena, was born, they moved to his home town where some of the artists were trying to form an academy. Another daughter, Dorothea Maria, was born to the young couple.

Between 1675 and 1680 Maria Sibylla produced a three-part book of illustrations of flowers, without text; some caterpillars, butterflies, spiders and other creatures were depicted on the plants, but otherwise it was a conventional book of its type. Then in 1679 she produced another, two-volume book, all about caterpillars, their wonderful transformation and singular flower-food. The insects were not merely adjuncts to paintings of flowers, they were there for themselves, shown in the various stages of their existence. The text gave a careful description of each stage; there were other books about insects being published at this time but hers was unique in giving a full account of their external appearance through the life cycle with illustrations all taken from life. Ten years previously the Dutch physician Jan Swammerdam had published his *General History of Insects*, the foundation of the new entomology. Beautifully produced, her book on caterpillars was published in Frankfurt in 1683. Linnaeus consulted many of her illustrations in the course of his work on a system of taxonomy, in some cases examining specimens she had prepared.

Two years later, against a background of family lawsuits, Maria Sibylla left her husband, resumed her maiden name and joined a religious community in Waltha Castle near Wieuwerd in West Friesland. This Protestant sect, the Labadists, had an international following, with some 350 adherents, both men and women, some of whom were elect, others just aspirants. They had to give all their possessions to the community and retire to live a spiritual life of love and discipline. She did not have to separate herself from her daughters and was allowed to continue her artistic and entomological work. After five or six years she changed her mind about the Labadists and left for the flourishing commercial city of Amsterdam.

Here she was welcomed by fellow naturalists and collectors: in the botanic garden she could see plants from the Americas, Africa and the Pacific, whose seeds or specimens had come from Dutch traders and officers of the Dutch East India Company. She could also visit the museum of anatomical and other rarities built up by professor Frederick Ruysch and the cabinet of curiosities formed by burgomeister Nicolas Witsen, president of the East India Company, whose tropical insect specimens she examined with wonder. She wanted to see for herself where these creatures came from and so 'I was moved,' she said, 'to make the long and costly journey to Suriname,' the Dutch colony in Guyana. The Labadists had tried, unsuccessfully, to establish an outpost there, but had been defeated by the climate.

Some descriptions of the flora and fauna of that part of the world had already been published by European settlers and government officials, but she had to defray her own expenses. To do so she sold a large collection of her paintings of fruit, plants and insects and specimens from the collection she had formed, as well as all her household effects. In 1699 she settled with her daughter Dorothea in Paramaribo, the chief town of the colony, and began work by making excursions into the interior, where she found the Africans and Amerindians more helpful to her than the European settlers. After almost two years she could not bear the heat any longer: 'I did not find,' she wrote, 'in that country a suitable opportunity to carry out the insect studies I had hoped to do, as the climate there is very hot. The heat caused me great problems, and thus I found myself compelled to return home sooner then I had anticipated.'

She returned, with her daughter, to Amsterdam in 1701, loaded with drawings and paintings, preserved butterflies, crocodiles and snakes, lizard's eggs, bulbs, chrysalises and pressed insects for sale. Four years later *The Metamorphoses of the Insects of Suriname* appeared in Amsterdam, a magnificent folio volume of sixty copperplates with text in Dutch and Latin. This was a success, widely read by naturalists, but still did not earn enough to cover the cost of production and pay back her travel loans. Her elder daughter Johanna Helena also went out to Suriname where she married the administrator of the orphanage in Paramaribo. The younger daughter Dorothea became the second wife of the Swiss painter Georg Gsell and went to live in St Petersburg; their daughter married the great Swiss mathematician Leonard Euler. After two years of ill health Maria Sibylla died on January 13, 1717.

SIR HANS SLOANE (1660–1753)

The Sloane family migrated to Ireland from Scotland during the reign of James I, and became wealthy. The subject of this profile was the youngest of seven sons born to Alexander Sloane and Sarah Hicks. Hans Sloane's first name was intended as a compliment to the Hamiltons, earls of Clanbrassill, a family in which it was common. Sloane's father was the earl's receiver-general of taxes from County Down. Nothing is known about the Hicks family except for a suggestion that Sarah's father may have been chaplain to Archbishop Laud.

In his youth Sloane turned his interest towards natural history: 'I had from my youth been very much pleas'd with the study of plants, and other parts of nature, and had seen most of those kinds of curiosities, which are

found either in the fields, or in the gardens or cabinets of the curious in these parts.' County Down, with Strangford Lough, presented many opportunities for the study of natural history. Sloane visited Copeland Island where he was much intrigued with the seaweed on the seashore, which the Irish were accustomed to chew in order to cure the disease of scurvy.

Sloane was probably consumptive. At the age of sixteen he was 'taken with spasms of blood . . . but avoided the consequences of a disorder which must otherwise have proved fatal to him'. Three years later in 1679 he was well enough to go to London to study medicine. He lodged near the laboratory of the Worshipful Society of Apothecaries, where he studied chemistry under Nicholas Staphorst and botany at the Apothecaries' physic garden at Chelsea. He attended lectures on anatomy and medicine, but most important at this period of his life were his friendships with two of the greatest English men of science of the day, Robert Boyle and John Ray.

In 1683 Sloane went to France. On his way to Paris he met the chemist Nicolas Lemery; in the French capital he visited the Jardin du Roi and frequented the Charité hospital. He also heard lectures on botany and anatomy. In those days it was impossible for a Protestant to take a university degree in Catholic France, but the Provençal town of Orange was still under the House of Orange. Its university gave examinations and conferred degrees but offered no instruction in medicine. Sloane graduated as doctor of physick there in 1683, then went to Montpellier to complete his studies, working under the physicians Charles Barbeyrac, Pierre Chirac and Pierre Magnol.

The persecution of Protestants in France was starting in 1684, when Sloane returned to London with the intention of practising medicine. For the contributions he had already made to botany he was elected Fellow of the Royal Society. The next year Robert Boyle recommended Sloane as a skilful anatomist to the famous surgeon Thomas Sydenham, who was not impressed by his academic studies and told him 'that is all mighty fine, but it won't do; . . . no, young man, all this is stuff; you must go to the bedside, it is there alone that you can learn disease'. The secret of Sydenham's success lay in his systematic approach to the symptoms presented by his patients, and Sloane followed his example. He was admitted a Fellow of the Royal College of Physicians of London in 1687, as so often at loggerheads with the Society of Apothecaries. One bone of contention was that physicians treated the poor free of charge but apothecaries charged for their medication.

When Christopher Monck, second duke of Albemarle, was appointed governor of Jamaica, he appointed Sloane to accompany him to the island as his personal physician. The expedition was of great value to Sloane, not only giving him first-hand experience of a relatively little-known island but also enabling him to search for new drugs. The description of the voyage and the observations on the inhabitants, diseases, plants, animals, some of which he brought back alive, and meteorology of the West Indies make Sloane's book on Jamaica valuable even today. The duchess of Albemarle was suffering from the first stages of the mental illness which turned to madness later. As a result the duke became an alcoholic and died within a year of arriving in Jamaica. Sloane escorted the duchess back to England in 1689.

Four years later he became secretary of the Royal Society, also physician to Christ's Hospital the following year. In 1695 Sloane married Elizabeth, daughter of John Langley and widow of Fulk Rose, formerly of Jamaica; they had four children, of whom two died in infancy. Their mother died in 1724. Sloane was now launched not only in the highest and scientifically distinguished society, but also in his profession of medicine, which became very lucrative. He might charge one guinea an hour to those who could afford it, nothing to the poor and needy. His wife's fortune, derived from her first husband's sugar plantation, made Sloane wealthy, apart from the income from his fashionable medical practice. In 1712 he was appointed physician to Queen Anne; four years later George I conferred a baronetcy on him, a kind of hereditary knighthood, but unlike a peerage not entitling the holder to a seat in the House of Lords. In 1712 he purchased the manor of Chelsea, treating the manor house as his country residence but did not make it his main residence until much later. He also purchased the adjacent mansion where Sir Thomas More had lived, then in a state of decay, but had it demolished. His Chelsea property included the historic physic garden, which he later conveyed to the Society of Apothecaries, on condition that they maintained it properly. Out of gratitude the society commissioned a statue of their benefactor which still stands in the garden. In 1715 he was elected president of the Royal College of Physicians.

A great believer in the importance of diet Sloane, who became familiar with chocolate in Jamaica, found it to be more digestible when mixed with milk. Sir Hans Sloane's milk chocolate was produced by Messrs Cadbury until 1885. He also played an important part in establishing the practice of inoculation against smallpox. In 1739 he was associated with Thomas Coram in the foundation of the Foundling Hospital, one of many philanthropic initiatives he supported. Having been secretary of the Royal Society since 1693 he succeeded Newton as president in 1727, serving until 1741 when he resigned owing to failing health. Sloane was a fellow of the society for almost sixty-eight years, which was a record. He entertained his scientific and medical friends regularly.

Throughout his life Sloane amassed collections. Beginning with botanical specimens collected in France and the West Indies, he added other collections of plants, animals, insects, fossils, minerals, precious stones and ethnographical specimens. He also branched out into Egyptian, Assyrian, Etruscan, Roman, Oriental, Amerindian and Peruvian antiquities. To these were added drawings and etchings by Albrecht Dürer, Hans Holbein and Wenzel Hollar, and a valuable collection of coins and medals. Sloane's library contained over 50000 books and 3500 bound volumes of manuscripts; his herbarium filled 337 folio volumes. Scholars had ready access to the collections, which were moved to Chelsea from Bloomsbury in 1742. The names of Sloane and his relatives are to be found all over his former Chelsea estate, notably in Sloane Street and Sloane Square.

In his will he offered all these treasures to the British nation, on condition that £20000 was paid to his daughters, who might have expected to inherit the valuable collection. After Sloane's death on January 11, 1753 the trustees whom he had appointed met; the matter was brought before parliament, which received Royal Assent for the Act which enabled Sloane's collection to be acquired and a suitable building purchased to house it. Thus the British Museum was founded; almost at once the Harleian collection of manuscripts and the Cottonian library were added, other collections later. The trustees used the parliamentary grant to purchase Montague House, in Bloomsbury, to accommodate them; this was where the mad duchess of Albemarle had lived out her many years, after marrying the extravagant earl of Montague, until the building was left empty and became derelict. The museum was theoretically opened to the public in 1759; admission was very restricted at first and visitors were hurried through by the staff without any time to study the exhibits. Montague House was pulled down and replaced by the more spacious present building in the 1840s, enabling hundreds of people to view the collections every day. Even so, there was nothing like enough space to exhibit all the contents of the museum, and in 1881 the natural history section, which contained the kernel of Sloane's collections, was moved to a new building in South Kensington, as we shall see.

ANTONY VAN LEEUWENHOEK (1632–1723)

The celebrated microscopist Antony van Leeuwenhoek was born in Delft on October 24, 1632, the fifth child of Philip Thonisz and Grieje Jacobsdr. van der Berch. His father was a basket-maker by trade, as was his own father before him; his mother came of a highly respectable although not particularly wealthy family. When he was five the boy lost his father; three years later he left home, and in the same year his widowed mother married again, this time to Jacob Molijn, 'city painter and bailiff of the common finances'. The boy received his early education at Warmond and then Benthuizen, where he had relatives. When he was about sixteen, he moved to Amsterdam where he was initially employed by a draper as book-keeper and cashier. Six years later he returned to his birthplace where he married Barbara de Mey in 1654. On marriage, Leeuwenhoek had bought a house in Delft, with a shop where he set up in business as a draper. Barbara died twelve years later, leaving him with Maria, the only one of their five children to survive childhood. In 1671 he married his second wife, Cornelia Swalmius. It is thought possible that he continued his education while in Amsterdam, where Cornelia's uncle, who was a physician, might have encouraged the interest in biology for which he was to become famous.

His long association with the Royal Society, of which he was elected a fellow in 1680, began in 1673 with letters of introduction written by Reinier van Graaf, a friend and fellow townsman and by the diplomat

Constantijn Huygens, father of the scientist Christiaan. Previously he had made a visit to England about 1668, probably to see Robert Hooke, who was about to publish his *Micrographia*, but otherwise he never left the Netherlands. Most of his discoveries were communicated in his famous correspondence with Oldenburg, the secretary of the Royal Society: his letters to the society were written in Nether Dutch, since he knew no other language. Some 120 extracts in English, or occasionally in Latin, were printed in the *Philosophical Transactions* between 1673 and 1723.

In 1668 he confirmed the discovery of blood capillaries by Marcello Malpighi (1628–1694), six years later gave an accurate description of the oval red blood corpuscles in fishes, frogs and birds, and the disc-shaped corpuscles in man and other mammals. The first written account of free-living ciliate protozoa appeared in a letter of 1674, and of bacteria found in water wherein pepper lain infused in a celebrated letter of 1676, which is of special interest to protozoologists. His finding of entozoic protozoa in the gut of the frog and of bacteria in the human mouth was described in 1683. Observations of the life history of rotifers, which he discovered, of fleas, aphis and ants and of marine and freshwater mussels confirmed his disbelief in spontaneous generation. Concerning this he wrote in 1702: 'Can there even now be people who still hold to the ancient belief that living creatures are generated out of corruption?' Among his miscellaneous discoveries were spermatozoa of dogs and other animals, hydra, volvox and the globular nature of yeast.

Given the prevalence of the mechanistic philosophy at the beginning of the eighteenth century it was very difficult to imagine ways in which new biological forms could arise from undifferentiated matter. To deal with this problem the preformationists went so far as to deny that there had ever been generation of any kind, whether of biological individuals or of members or parts. Thus when Leeuwenhoek peered through his microscope at semen, he fancied he saw a tiny foetus at the head of every sperm. When properly planted in the female womb these tiny creatures would enlarge into infants.

Leeuwenhoek was also one of the first to study the structure of opaque objects by means of sections cut by hand with a sharp razor. His researches followed no scientific plan and were made, for the most part, with the aid of microscopes made by himself. These consisted of a single very small biconvex magnifying glass of remarkable clarity, mounted between small apertures in two thin oblong metal plates, usually of brass riveted together. The instrument was held close to the eye, and the object, mounted on a silver needle on the other side of the lens, was adjusted to correct focus by means of a thumbscrew. Compound microscopes of that period were no use for scientific research. He never sold any of his instruments, nor was he at all generous in giving them away. He never had any assistants or pupils, and it is something of a mystery how he obtained his extraordinary results.

During his lifetime Leeuwenhoek became famous, and corresponded with such savants as Leibniz, the younger Huygens and Magliabecchi. Crowned heads and princes came to pay homage as well as eminent scientific visitors with whom he was in correspondence. In fact he had so many visitors – twenty-six in four days – that he had to send some away. Yet we know very little about the man, apart from his work. Apart from his six years in Amsterdam as a young man, he lived all his adult life in Delft, where he was born. It is thought possible that he gave up shopkeeping after inheriting some money from his mother's family. He served as Chamberlain to the Sheriffs of Delft from 1660 to 1699. In 1676 he was appointed trustee of the insolvent estate of the deceased painter Jan Vermeer, about whom we also know frustratingly little, but whose paintings give us some idea of the world in which Leeuwenhoek lived. The various civic offices he held brought him a small income and he was at least comfortably off later in life. He died on August 26, 1723; his second wife had died almost thirty years before. He did not marry again but was looked after faithfully by Maria, the surviving daughter by his first wife.

Leeuwenhoek left 274 finished microscopes and 172 lenses he had made. He bequeathed to the Royal Society a cabinet containing twenty-six instruments and extra lenses, which was lost sometime in the nineteenth century. Before this the optical properties of the microscopes were ascertained; they were found to have magnifying powers ranging up to 200; one in the museum at Leiden has a power of 270. Since he kept his method for seeing the smallest animalcules to himself alone, there has been speculation as to what his closely guarded secret could have been. Dobell (1932) believes that Leeuwenhoek discovered some form of dark-ground illumination, and then found that the effectiveness of lenses similar to those employed by Leeuwenhoek can be augmented by utilising the optical properties of spherical drops of fluid containing the objects under observation. Among his contemporaries there were others, notably his compatriot Jan Swammerdam (1637–1680), who also used the early microscope with extraordinary skill, but it was Leeuwenhoek who opened the eyes of his fellow biologists to the existence of the teeming world of the microscopic.

2 From Réaumur to Hunter

RENÉ-ANTOINE FERCHAULT DE RÉAUMUR (1683–1757)
Réaumur was among the greatest naturalists of his own or any age in the breadth and range of his researches, in the patient detail of his observations, and in the brilliant ingenuity of his experiments; Huxley compared him favourably with Darwin. He was of an illustrious Vendée family, the Ferchaults, who prospered in commerce and purchased the ancient Réaumur estate early in the seventeenth century. Through a protracted lawsuit Réaumur's grandfather, Jean Ferchault, obtained half the seignorial rights over his newly acquired fiefdom and thus entered the ranks of the lesser French nobility. René Ferchault, Réaumur's father, was a conseiller au présidial at La Rochelle, a position corresponding to an appellate judge in an intermediate provincial court. He married Geneviève Bouchel, the youngest daughter of one of the leading citizens of Calais, in April 1682; René-Antoine was born the following February 28. Réaumur's father died nineteen months later. A second son, Jean-Honoré, was born posthumously. Réaumur and his brother were raised by their mother with the aid of several aunts and uncles.

Concerning Réaumur's early education nothing is known with certainty. Probably he studied with either the Oratorians or the Jesuits at La Rochelle. Then, in accordance with established custom among the bourgeoisie and lesser nobility of the region, he would most likely have been sent to study with the Jesuits at Poitiers. In any case Réaumur's early education is unlikely to have included much physics or mathematics. In 1699 his maternal uncle, Gabriel Bouchel, summoned him to Bourges to study law. He went there with his younger brother and stayed three years. In 1703 Réaumur went to live in Paris, where he met a cousin on his mother's side, Jean-François Henault, later to be president of the Parlement in Paris. Henault was studying mathematics with a certain Guisnée, an obscure 'student geometer' of the Académie des Sciences (geometer meant mathematician in those days). Réaumur decided to take lessons from Guisnée and, according to Henault, after only three sessions knew more than his cousin and as much as his instructor. It was probably through Guisnée that

René Antoine Ferchault de Réaumur. Commandeur et Intendant de l'Ordre royal et militaire de St. Louis.

Réaumur became acquainted with the academician Pierre Varignon, professor of mathematics at the Collège Mazarin and at the Collège Royal. Varignon also gave Réaumur private lessons in mathematics and physics, and became not only his teacher but also his friend and guide. In 1708 Varignon nominated Réaumur to be his 'student geometer' at the Académie des Sciences.

Guided by Varignon, Réaumur then devoted himself to research in pure mathematics. His first three communications to the academy, on geometrical subjects, were presented in 1708 and 1709, and demonstrate such a degree of mathematical sophistication that Varignon thought that Réaumur had the potential to become one of the greatest geometers of his age. In November 1709, however, Réaumur quite suddenly changed to natural history after reading a paper on the growth of mollusc shells, which maintained this was due to the incorporation of new matter into an already existing structure. He showed that, on the contrary, it was due to the addition of successive layers.

The Paris Academy, unlike the English Royal Society, was part of the French apparatus of government. Its official role became more and more pronounced throughout the eighteenth century as academicians assumed administrative control of French technology as consultants, inspectors and even directors of industry. Shortly after its foundation Colbert, the finance minister of King Louis XIV, charged it with the task of providing a description of all the arts, industries and professions. This was intended to form a sort of industrial encyclopaedia in which the secret processes of industrial technology would be described and, where appropriate, improved. Réaumur was one of the earliest and most enthusiastic supporters of this technocratic function of the academy, and perhaps it was for this reason that he was given the task of producing the encyclopaedia that Colbert wished to see. After 1713 this took up much of his time and effort, at the expense of research into natural history.

Although he also examined the tinplate and porcelain industries, his most significant and original contribution to industrial technology was an examination of the processes used in the production of iron and steel, the results of which he presented in a series of memoirs to the academy in 1720–1722. The regent Philippe II, duke of Orleans, took a keen interest in this and helped him obtain information about the ferrous metals industries in other countries. The industry was technologically backward in France and it was hoped that Réaumur's report would help to remedy the situation. In 1715 he turned his attention to the production of artificial pearls and this led him to investigate how the scales of fish gained their lustre. He attempted to stimulate pearl formation artificially. Next he rediscovered the secret of making the purple dye of the ancient Romans from the substance produced by a particular species of mollusc. He also investigated the means by which molluscs, starfish and other species of invertebrate move around.

In natural history Réaumur's most important work was his *Mémoires pour servir a l'étude des insectes.* The term 'insect' was used in the early eighteenth century to designate almost any small creatures possessing segmented bodies; thus spiders and worms were usually included and for Réaumur also polyps, molluscs, crustaceans, and even reptiles and amphibians. It is difficult to know at times just what exactly he is referring to. He was fascinated by the behaviour of insects, by their industry, diligence, ingenuity, organisation and skill. His original plan had been to write ten of these memoirs; six were published in his lifetime but after his death nothing was found but fragments in manuscript, some of which were not published until quite recently. There is some evidence that he may have been discouraged from continuing the series by competition from his younger and more popular contemporary Buffon, the subject of the next profile.

Réaumur was particularly interested in bees, where his research was both extensive and original. He kept track of individual bees by tinting them with various dyes. He dissected bees and their larvae and had detailed plates made to illustrate his memoir. He discovered that all hives, even those very close to swarming, have only one queen; if others are introduced they will be rejected or even killed; he found that if a colony is deprived of its queen, it will make a new one by feeding a special substance called royal jelly to the developing larva. He discovered that a hive without a queen will under certain conditions accept an exogenous ruler. He also made some of the first tentative studies of communication among bees.

Réaumur was in touch with a cousin of his named Abraham Trembley, who had studied philosophy and mathematics in Geneva. Trembley read the *Mémoires des insectes* with great interest and in 1740 passed on to its author his observations on freshwater hydra or polyps. Were these animals or plants? Réamur thought the former, since they had animal-like powers of locomotion, expansion or contraction. But when a polyp was cut into pieces each regenerated into a whole polyp. The observation of analogous regenerative powers in starfish, sea anemones and worms was likewise highly disturbing to the commonly received notions of biology. Trembley wrote an important monograph on the subject, published in 1744.

Réaumur is perhaps best known for the thermometric scale that bears his name. Thermometers had been used for about a century and the Fahrenheit scale was beginning to be adopted in both England and the Netherlands but it was accurate only if the inside of the hollow thermometer tube was perfectly regular. Réaumur overcame this problem but created others so that his scale was never widely adopted. He was perhaps the most prestigious member of the Académie des Sciences during the first half of the eighteenth century; through his incessant labours and voluminous publications, through his extensive correspondence with scientists at home and abroad, and through the reflected brilliance of his students, he acquired enormous authority and renown in the European scientific community. Between 1713 and 1752 he was appointed deputy director of the academy ten times and director nine times. He was elected to the Royal Society of London, to the scientific academies of Prussia, Russia and Sweden, and to the Institute of Bologna.

Réaumur never married but devoted all his time to his scientific and academic career. From a needy relative he purchased the title of commander and intendant of the Royal Military Order of Saint Louis, an honorific office possessing the dignity of count. Two years before his death he inherited from an old friend the castle and lordship of La Bermondière in the Maine. The castle and surrounding land had been turned into a fine eighteenth-century chateau set in beautiful grounds. He was staying there when one day he suffered a fall from his horse and died the following night, on October 19, 1757.

GEORGES-LOUIS LECLERC, COMTE DE BUFFON (1707–1788)

Georges-Louis Leclerc, who later became comte de Buffon, was born on September 7, 1707 in Montbard, Burgundy, the eldest son of Benjamin-François Leclerc and Anne-Christine Marlin. His father was a lawyer, the son of a judge, who was himself the son of a doctor. His mother brought with her a generous dowry, which enabled the Leclercs to move to Dijon, the provincial capital. Their son attended the Jesuit-run Collège de Godrans, which had a good reputation. Although the curriculum was the usual humanist one, as other Jesuit schools it included regular classes in mathematics and physics. At this stage he became particularly interested in mathematics. At the age of sixteen he enrolled in the local law school, leaving it three years later after having decided against a career in the law. Instead he would dedicate himself to science.

He left Dijon at the age of twenty to study mathematics in Angers. After fighting a duel he had to return to Dijon where he became friendly with a pleasure-loving young English aristocrat, the duke of Kingston, who was on the customary extravaganza known as the Grand Tour. They went off, together with the duke's tutor, to Lyons, Geneva, Turin, Milan, Genoa, Pisa, Florence and Rome, mainly having a good time. After they separated, Leclerc returned to Dijon, where his recently widowed father had remarried. There was a dispute between father and son over the settlement from the previous marriage, which should have come to the son. Their relationship was strained for a long while afterwards. In 1732 Leclerc moved to Paris, at the age of twenty-five, to launch himself on a career in science. After he wrote two mathematical papers, which were accepted by the Académie Royale des Sciences, he was elected to the lowest grade in the academy in 1736.

In Burgundy he now replaced the modest house where he was born by an imposing chateau set in parkland, using the ruins of a feudal castle to

provide the necessary building material. For the next fifty years he spent the summers on his estate, the rest of the year in Paris. Although by no means unsociable, his acquaintances were mainly intellectuals like himself; he did not much care for the social life of the French capital which revolved around the famous salons. He established contacts with English scholars, in whose scientific work he was greatly interested, and in 1739 was elected a member of the Royal Society of London.

In the eighteenth century French universities, particularly the Sorbonne, were in decline. To compensate for this Richelieu had persuaded Louis XIV to found several institutions, one of which was the Jardin du Roi, originally a physic garden consisting of medicinal plants. This was administered by an intendant; when this post became vacant in 1739 Leclerc was appointed. This was considered a surprising choice but Leclerc had transferred from the mechanical to the botanical section of the academy because he had become more interested in chemistry and biology, especially forestry, than mathematics. Thus he had some professional qualifications for the post but, in this age of patronage, what mattered more was that he knew someone who had influence and would press his case. During his long tenure of the post at the garden, where he enjoyed an official residence, the collections were greatly expanded. Leclerc's career was focused on the Jardin, not the Académie des Sciences; in its government his authority was paramount.

However Leclerc is known to posterity mainly for his encyclopaedic *Histoire naturelle* in thirty-six volumes. The full title of Leclerc's great work, natural history, general and particular, indicated its comprehensive nature. Critics agreed that it was beautifully written, although what he wrote was often controversial. Scientists in the eighteenth and nineteenth centuries were obliged to try and reconcile scientific knowledge with religious teaching. It was not necessary, or even possible, to take literally the account of the creation of the world from the Book of Genesis. Some reinterpretation was acceptable, and Leclerc gave what was current in scientific circles at his period. However the origin of species was another matter. Were the different species of animals created separately, or did they have a common origin from which they had diverged? In Leclerc's time this fundamental question was beginning to be debated but was a long way from being answered. His *Histoire naturelle* covered almost everything that could be written about the subject at that time in a popular work. Dedicated to the king it was sumptuously produced, with numerous illustrations. After complimentary copies had been distributed in 1749 it was placed on general sale, but was not a commercial success. The work made him famous internationally: foreign academies showered him with honours. In France he was elected to the literary Académie Française, and before long was director of that august body.

In 1752 Leclerc had married the pretty twenty-year-old Françoise de Saint-Belin-Malain. He chose a provincial bride of noble birth, from the Ursuline convent in Montbard of which his sister was Mother Superior. She was not yet a nun but would have no doubt become one if she had not married. She proved a good hostess, entertaining the many guests who came to stay. Occasionally she visited her husband in Paris but most of her time was spent at the chateau. In 1758 she gave birth to a daughter who survived barely eighteen months. In 1766 Françoise injured herself by falling from a horse, and never fully recovered. She died in 1769 at the age of thirty-seven, leaving a five-year-old son, George-Louis-Marie.

The grant of reversions to official positions was common practice under the *ancien régime*, even for positions where a scientific or even medical qualification would seem an essential requirement. To provide for his son Leclerc obtained from Louis XV the reversion of his position as intendant of the Jardin du Roi. He also arranged that his son should receive two-thirds of his own salary. Leclerc fell seriously ill in 1771, perhaps from an abscess in the intestines, and his life was thought to be in danger. However his son was only seven years old at this time and would not have been able to take up the position until he was twenty-five. The king transferred the reversion to someone else, with Leclerc's agreement, and it was probably this which led the king to raise him to the nobility as comte de Buffon. He also ordered a statue to be made of Buffon, paid for out of his personal privy purse. Finished in 1776 it can still be seen in the Jardin. Several busts of Buffon by different sculptors, including Houdon, were made later.

As a member of the noblesse de robe Buffon was exempt from taxes himself but his seignorial rights included the tierce, a levy on the produce of the farmers of his county. In return he was expected to provide them with amenities but, like others in his position, did not regard this as a duty. To help cover the running expenses of his chateau he obtained some income from the lands of his estate, especially the forests, where his scientific knowledge came in useful. He also hoped to profit from a blast furnace with two forges attached which he designed and had built but this was not a success. In Paris the expansion of the Jardin involved the purchase of some land; he bought this himself and then resold to the crown at a much higher price. This kind of abuse was quite normal under the *ancien régime.*

After the death of Françoise, Buffon installed a young peasant girl named Marie Blessau as housekeeper at Montbard. Devoted to her master she looked after his material needs for the rest of his life. Other young women enlivened the social life of the chateau. At this stage in his life he preferred to dress in an old-fashioned style, and reproved his son for not following his example. He was always well groomed, having his hair curled every day, sometimes several times. He got up very early in the morning, took a light breakfast and went to work in his office until dinner was served at two o'clock. One of his guests left this impression of a visit in 1785, when the great man was seventy-eight.

> It is at dinner that he places his mind and his genius aside; there he abandons himself to all gaieties, to all follies that go through his head. His greatest pleasure is to make naughty remarks, which is all the more agreeable because he retains his characteristic calm; his laugh and his old age form a sharp contrast to the seriousness and gravity that are natural to him, and these jokes are often so shocking that the ladies are forced to leave the room. After dinner, he hardly bothers with those who are staying at the chateau or the others who have come to see him. He goes to sleep for half an hour in his room, then he takes a walk, always alone, and at five o'clock he returns to his office to go back to his studies until seven o'clock; then he returns to the salon, has his works read, explains them and admires them.

On marriage Buffon had moved out of the official residence to a mansion more in the centre of Paris. He moved back again in 1772 to a house near the Jardin, where he enriched its collections and enlarged its buildings and lands considerably but in a high-handed way that made him enemies. In the spring of 1788 he was taken seriously ill and died on April 16, not long before the Revolution started. At the Jardin he was succeeded by two intendants of markedly inferior scientific reputation and personal authority. Its name was changed to the Jardin des Plantes while the building housing the collections became the Muséum d'histoire naturelle.

Buffon's son, a spoiled child, never had much to do with his father. He lived in Paris, neglecting his studies. At the age of sixteen he enlisted in the Guards. It was decided he should see something of the foreign lands by accompanying the royal botanist Lamarck on an official tour of botanical gardens and museums in Austria, Hungary and Germany. After complaints from Lamarck about his conduct, Buffon recalled his son to Paris. The next year he set off again, this time to Russia, with an officer of the Guards as companion. In 1783 he was named governor of Montbard. The following year he married the ravishing sixteen-year-old Marguerite-Françoise, daughter of the late marqis de Cepoy. She brought a handsome dowry which, with the allowance provided by the count, made the young couple wealthy. Unfortunately the young lady was already being pursued by Louis-Philippe, duke of Orleans and the liaison became public knowledge. She left her husband and became the duke's official mistress. In 1793 the comte de Buffon, as he had become after the death of his father, divorced her and married a childhood friend. Although he supported the Revolution the following year he was condemned to death and died on the guillotine during the Terror.

CARL LINNAEUS (1707–1778)

Nil Ingermarsson Linnaeus, a country parson in the poor and remote south Swedish province of Småland, settled in Stenbrohult, where his son Carl, the future botanist, was born on May 23, 1707 and grew up. The mother, Christina Brodersonia, was a parson's daughter. The father was a great lover of flowers and laid out a beautiful garden in the parsonage grounds. He also introduced his son to the mysteries of botany. In 1716 Linnaeus entered the grammar school in the nearby cathedral city of Växjö, and from that time natural history remained his favourite study. An average boy, he was set apart only by his delight in botanising and learning about plants. He was fortunate in having an outstanding teacher during his last years at school, who encouraged his aptitude for botany and taught him the classification

system of the illustrious French botanist Joseph Piton de Tournefort (1656–1708); this was of great importance for his future development

In 1727 Linnaeus entered the University of Lund to study medicine. The medical faculty was rather mediocre, but during his year at Lund Linnaeus made several extensive botanical field trips in the surrounding area. He was taken up by the scholarly medical doctor Kilian Stobaeus, one of a number of patrons who helped him in his youth. Stobaeus let Linnaeus use his rich library and collection of botanical specimens. In the autumn of 1728 Linnaeus transferred to the more important University of Uppsala, although the medical faculty was not much better than at Lund. He found the botanical garden at Uppsala, although somewhat neglected, contained rare foreign plants. Uppsala also provided him with influential patrons: the elderly professor of medicine Olof Rudbeck the younger, formerly an outstanding botanist and zoologist, and the learned Olof Celsius, with whom Linnaeus studied the flora of the surrounding region.

Around 1730 Linnaeus began to conceive the fundamental features of his botanical taxonomy, based on plant sexuality, an idea originally proposed by the French botanist Sebastien Vaillant. He barely supported himself during this period of poverty by substituting for Rudbeck in conducting demonstrations at the botanical garden and by giving private tuition. In the spring of 1732, sponsored by the Uppsala Society of Science, he made a trip to Lapland that lasted until the autumn. His account of this journey gives vivid evidence of his joy in the unfamiliar world of the Arctic tundra, where he studied not only the unknown alpine flora but also the customs of the Sami reindeer herders. Two years later he undertook another, less ambitious, field trip to the Dalarna region of central Sweden, and then went abroad for the first time.

In the spring of 1735 Linnaeus moved to Holland for three years, where learned and wealthy patrons helped him get his work published. He based himself at the University of Leiden, one of the best in Europe at that time, but he also made a short visit to England, where he stayed with the wealthy merchant George Clifford, whose botanical garden at the Hartekamp, near Haarlem, was famous. In London he met Sir Hans Sloane, John Dillenius and other naturalists and was impressed by the apothecaries' physic garden in Chelsea. He returned to Sweden in 1738 via Paris, where he met the brothers Jussieu. Although he had already established his reputation as a leading young botanist there was no academic position available for him. However he had already obtained his medical qualification, with a lightweight thesis on the causes of malaria, and so he could begin to practise in Stockholm. The same year he married Elisabeth Moraea, the daughter of a wealthy physician of the city. Within a year, in 1739 he was appointed chief physician to the Swedish Navy, specialising in the treatment of syphilis. He was one of the prime movers in the foundation of the Swedish Academy of Science, of which he became the first president. Later he was elected a Fellow of the Royal Society.

In 1742, at the age of thirty-five, he was appointed professor of medicine and botany at the University of Uppsala, where he remained for the rest of his life. He moved in to the official residence in a corner of the botanical garden, for which he was responsible. Students loved his lectures, which were characterised by humour and the presentation of unusual ideas. They covered the whole of natural history, especially botany, and were supplemented by field trips in the country around Uppsala. He organised these on military lines; on the return home the participants marched in closed formation to the music of hunting horns and kettledrums. These trips were immensely popular, so much so that six years later the rector of the university curtailed them, explaining to Linnaeus 'we Swedes are a serious and slow-witted people; we cannot, like others, unite the pleasurable and fun with the serious and useful'.

Unable to accept criticism, Linnaeus sulked like a child whenever he encountered it. After this incident, which upset him greatly, Linnaeus started to reform Tournefort's taxonomic system. He devised the binomial system of classification which bears his name, assigning to every known organism a genus and a species. He used Latin, still the scientific lingua franca; this had often been used for concise scientific descriptions of plants and animals, but not in a system of classification. His method of classifying plants was based on a single characteristic, the sexual organs. Seeking simple ways for professional and amateur botanists to identify plants, Linnaeus aimed for the practical, recognising that the flower and the fruit are relatively stable characters, not subject to the great variation seen in other parts of plants. The characters that Linnaeus used for his classification system included the number of pistils and stamens, the presence or absence of flowers, and the presence or absence of both sexual organs on the same plant. Most naturalists followed his methods of nomenclature and description for taxonomic work well into the nineteenth century, but since it was not always based on natural relationships, modifications of the original system were preferred. He also extended his classification to the animal kingdom, but with less success.

Linnaeus gave his contemporaries the means of identifying the plants and animals then known, and thereby stimulated further collecting and exploration by his correspondents and students. He corresponded with the world's botanists and collectors who supplied him with plants for his herbarium, which became very extensive. In his mature years Linnaeus made official visits to the various Swedish provinces, to provide the Swedish government with an inventory of the natural resources of the country. From 1759 he was assisted by his son Carl the younger, who had been appointed botanical demonstrator and who eventually succeeded his father in his professorial chair at Uppsala.

Linnaeus had an unerring instinct for useful social connections; he counted among his friends and admirers the country's outstanding cultural leaders. In Sweden he was appointed court physician in 1747 and fifteen years later was elevated to the nobility, after which he was known as Carl von Linné (the name Linnaeus had been added to his patronymic by his father; it refers to the linden tree which grew in the garden of the parsonage). Linnaeus occasionally worked for the royal court; he described the valuable natural history collections at the castle of Drottningholm, and charmed the discerning queen, Lovisa Ulrika, by his unaffected manner.

In later years he travelled every summer to Hammarby, his small estate outside Uppsala, where he devoted himself to botany and teaching a few selected students. Linnaeus took an unselfish and devoted interest in his more mature students, supporting and encouraging them. Some he sent on long-distance voyages, for example Pehr Osbeck. The son of landless peasants, Osbeck won a scholarship to the university where he studied under Linnaeus. In 1750 his teacher secured for him the post of chaplain on a ship bound for Canton. He collected foreign plants en route and tried to sort them out on his return and it was the struggles of people like Osbeck that seem to have stimulated Linnaeus to devise his practical system. Another he inspired was Carl Peter Thunberg (1743–1828), who set off from Sweden for a journey through Japan, Java, the Cape and Ceylon, where he found two thousand new species of plants.

Linnaeus himself was very egotistical: he considered his published works to be unblemished masterpieces, no one had ever been a greater botanist or zoologist. Most of his students worshipped him but others found his personality repellent and over the years he became increasingly isolated. Of unstable temperament and essentially naïve, he vacillated between overweening pride and brooding despair. In a sunny mood his charisma was irresistible. Yet he was also a rude provincial – sentimental, superstitious and devoid of general culture, possessing the thrifty, tenacious, enterprising character traditionally ascribed in Sweden to natives of Småland. After Linnaeus suffered a fit of apoplexy in the spring of 1774 his mind became increasingly clouded; another fit followed a fall in 1776, and he died on January 10, 1778. His widow sold his collections and library to Sir James Smith of Norwich; the founder and first president of the Linnean Society of London, which purchased his botanical specimens, library and correspondence from Smith's widow in 1829 and has treasured them ever since.

JOHN HUNTER (1728–1793)

As an investigator, original thinker and stimulator of thought, John Hunter stands at the head of British surgeons. He was an all-round naturalist but his great achievement was to have discovered the scientific principles of surgery. He was born in the Lanarkshire village of Long Calderwood, a few miles west of Glasgow, on February 13, 1728. He was the youngest of ten children, including the brilliant future anatomist William Hunter (1718–1783). His father, of the same name, was said to be a man of intelligence, integrity and 'anxious temperament'; his mother Agnes Paul, daughter of the treasurer of the city of Glasgow, 'an excellent and handsome woman'. John's education was neglected; he was probably dyslexic.

After a spell in Glasgow, staying with one of his married sisters, at twenty years of age he moved to London and began to assist his Paris-trained

elder brother who was prospering as a teacher and as a surgeon. In 1754 John entered St George's Hospital as a surgeon's pupil, while continuing to act as William's assistant. One of his functions was to deal with the 'resurrection men', who surreptitiously supplied bodies to be dissected in anatomy lessons. He was said to have been very popular with these men, and to have enjoyed such aspects of London life as the playhouses. William helped John to get launched on a career in surgery, training at St George's Hospital, first as a surgeon's pupil, then temporarily as house surgeon. For a short time he studied at Oxford, as a member of St Mary's Hall (later taken over by Oriel College), but concluded that the university was no use to an aspiring surgeon.

The next few years he was back assisting his brother but after a series of arguments about priority of discoveries, they went their separate ways. England and France, as so often, were at war. In 1760 John served as staff surgeon with the ill-fated expedition to Belle Isle and later with the army in Portugal, acquiring some valuable experience in dealing with gunshot wounds. In 1763 he returned to London and began to practise as a surgeon, while running a private class for anatomy and surgery. Among the famous men who trained under him was the great Edward Jenner. He obtained the first refusal of animals that died in the menagerie of the Tower of London. The next year he purchased two acres of land at Earls Court, where he had a house built where he could experiment with surgical technique and keep some of the rare creatures that travellers used to send him. These animals were not kept in cages, as in a zoo, but wandered freely around the grounds. Often he drove into London in a carriage drawn by a troika of buffaloes.

In research he was well ahead of William, by this time, and in 1767 he was elected a fellow of the Royal Society, where he often lectured in later years, although he never found lecturing easy. In 1768 he became surgeon to St George's, and in 1776 surgeon to King George III. He married Anne Home, a minor poetess, in 1771; she had a taste for fashionable society which he did not share. They lived in Jermyn Street until the end of 1783 when he bought the lease of two houses, one on Leicester Square, the other on Castle Street, with land that connected them. During the next two years he built, at great expense, a museum on the intervening ground, in which he exhibited his extensive anatomical collections. The arrangement is said to have suggested to Robert Louis Stevenson his story *The Strange Case of Dr Jekyll and Mr Hyde*, since Hunter conducted his surgical school at the Leicester Square end, and received cadavers for dissection at the other. Hunter suffered from angina pectoris, and it was an attack following a confrontation with his surgical colleagues at St George's which led to his death on October 16, 1793. In 1859 his remains were moved to Westminster Abbey. He left four undistinguished children, of whom two survived him, but Hunter's most lasting monument is his museum. He early envisaged the formation of a collection to illustrate comparatively the structure and function of organs throughout the animal kingdom. To this end he spent many hours daily dissecting and preparing his exhibits, assisted by a young Cornishman named William Home Clift, whom he trained in comparative anatomy.

After Hunter's death his museum was lovingly cared for by the admirable Clift until it was purchased by the nation for £15 000, and entrusted to the newly founded Royal College of Surgeons, under the super-vision of a distinguished board of Hunterian trustees, who moved it to Lincoln's Inn Fields and appointed Clift as conservator. This very famous collection suffered serious damage through enemy action on May 10, 1941, but has recently reopened on a smaller scale. Hunter's brother William, the successful physician, anatomist, collector and educator, also founded a

Hunterian museum. He had become extremely wealthy, through his practice as a gynaecologist, and assembled an enormous hoard of coins, minerals, scientific instruments, ethnographic items and works of art. When he died, ten years before his brother, he willed his collection to the city of Glasgow.

In person we are told that John Hunter was of middle height, vigorous and robust, with high shoulders and rather short neck. His features were strongly marked, with prominent eyebrows, pyramidal forehead and eyes of light blue or grey. His hair in youth was a reddish yellow, which whitened with age, as can be seen in his famous portrait by Sir Joshua Reynolds. In manners Hunter was impatient, blunt and unceremonious, often rude and overbearing, but he was candid and unreserved to a fault. He was usually taciturn, but when he spoke his words were well chosen, forcible and pointed, often broadly or coarsely humorous. But although he could never spell well or write grammatically, and his writings were carefully revised by others before they were printed, they preserve his ruggedness of style. His daily routine was to rise at five or six to dissect, take breakfast at nine, see patients till twelve, and visit his hospital till four, when he dined. He slept for an hour after dinner, then read or prepared his lectures, made experiments and dictated the results of his dissections. He was most punctual and orderly in his visits to patients, leaving a duplicate of his visiting-book at home, so that he could be found at any time.

3 From Banks to Humboldt

SIR JOSEPH BANKS (1743–1820)

Sir Joseph Banks dominated the scientific scene in Britain for roughly half a century. He was born in London on February 2, 1743, the only son of William and Sarah Banks of Revesby Abbey in Lincolnshire. From his father, who died in 1761, Banks inherited an ample fortune and the Revesby estate, which he resolved to spend on promoting his scientific interests when he came of age. From his mother, who lived until 1804, he seems to have acquired in childhood a taste for natural history. After some years with a private tutor he went to school at Eton College and then to Christ Church, Oxford. At the university his early interest in botany grew stronger but because teaching in the subject was virtually non-existent he brought Israel Lyons over from Cambridge to be his mentor. On leaving in 1763, without having taken a degree, he persuaded his mother to rent a house in Chelsea, to be near the physic garden. The same year, anxious to extend his knowledge of plants beyond those native to his homeland, he sailed as a scientific observer on HMS *Niger*, a fishery protection vessel voyaging to Labrador and Newfoundland.

After his return he came to know the botanist Daniel Solander, the much-loved pupil of Linnaeus, who was employed as an assistant librarian in the British Museum, and they became firm friends. Banks learnt that the Royal Society, to whose fellowship he had just been elected, was planning an expedition to the South Pacific to observe the transit of the planet Venus across the Sun in 1769. Immediately, he realised the opportunities this would offer for botanical work in lands whose rich flora and fauna were then virtually unknown. In due course the expedition, under the command of Captain James Cook, sailed in the *Endeavour* on what is now recognised as one of the great voyages of exploration and observation. Banks, accompanied by a party of seven – the chief of whom was Solander – sailed with him; the voyage is said to have cost Banks £10 000 from his own pocket.

The *Endeavour* left Plymouth in late August 1768 and reached Rio de Janeiro in mid November. There the Portuguese officials, suspicious of their intentions, tried to prevent them from landing, and there was not

much opportunity for botanising. They rounded Cape Horn the following April and reached Tahiti in good time to observe the transit of Venus which took place on June 3. Their next port of call was New Zealand, where six months were spent in exploration. After this they went to Australia, found an abundance of interesting plants in what they decided to call Botany Bay, and then sailed north inside the Great Barrier Reef of north-east Queensland. They were delayed for several weeks when the *Endeavour* struck a reef, then sailed on past New Guinea to Batavia, a Dutch colony on the Malayan coast, where almost everyone on board was stricken by fever. Seven crew members died while they were there, sixteen more later, including two members of Banks' party. They sailed home via the Cape of Good Hope and landed at Deal in mid June, after a historic circumnavigation.

The journals of his early travels are the record of a young man of great curiosity with a zest for discovering the new and the curious. Not only did he search enthusiastically for hitherto unrecorded flora and fauna but he also took an interest in the human societies he encountered in his travels. This anthropological interest was especially noticeable in Tahiti where during the *Endeavour*'s stay he was able to get to know something both of the people and of their language; he even had his body tattooed in the Tahitian manner. Banks returned, after an adventurous voyage lasting three years, with a vast quantity of material, to find himself a celebrity. He was summoned to Windsor to give a personal account of his experiences to King George III; this marked the beginning of a lifelong friendship.

In 1772 he prepared to make another voyage, with Cook, in the *Resolution*, but at a late stage he withdrew. Apparently this was because a laboratory, which he had ordered to be constructed on the deck, had been demolished by Cook who feared it might destabilize the ship. The young Banks was a man very used to having his own way. One who was present recorded that when he saw the ship and the alterations that were made he swore and stamped on the wharf like a madman and instantly ordered his servants and all his things out of the ship. Instead he made a voyage to Iceland, while the émigré German naturalist Johann Reinhold Forster, with his son George, took his place on the *Resolution*. Banks saw himself as a pioneer in linking naval voyages of exploration with scientific discovery and maintained a most active interest in such voyages, although he never again participated in one himself.

Banks' influence in high places was certainly a factor in his election to the presidency of the Royal Society in 1778, which he regarded as the greatest of the many honours that befell him in his life. As president he was determined to dominate, rejecting interested amateurs from the leisured classes as candidates for election unless he considered they might be useful as patrons of science. He also displayed a preference for natural scientists over physical scientists, whereas under Newton and his immediate successors it had been the other way round. Moreover the secretaries had become accustomed to act without presidential approval, and he had a major struggle to put this right. The society was at odds with the sovereign over the absurd controversy as to whether the tips of lightning conductors should be round or pointed and Banks was anxious to re-establish good relations.

At that time the Royal Society was housed in cramped accommodation in Somerset House, and Banks' own residence, in Soho Square, which he purchased in 1776 and where he entertained lavishly, quickly became the focus of the scientific world of London. Some of his contemporaries called it an Academy of Natural History. There scientists could not only meet their colleagues in agreeable circumstances but, no less important, could also meet leading figures from other walks of life. He was a man whose company others generally enjoyed, among them Samuel Johnson. When Johnson introduced Banks to Boswell the latter described him as an agreeable young man, easy and communicative, without any affectation or appearance of assuming. Nor was Banks one to bear grudges. With its rich library and herbarium, the house in Soho Square became the botanical centre not simply of London but of Britain. His younger sister, Sarah Sophia, came to live at the house and acted as his companion, hostess and amanuensis. Solander was his resident librarian until his death in 1782, after that first Jonas Drysander and then, from 1819 to 1820, Robert Brown, whose profile follows later.

At the age of thirty-six Banks gave up the life of a wealthy bachelor, in a society which tolerated occasional liaisons with the opposite sex, and married the much younger Dorothea Weston-Hugessen, an heiress who brought with her an estate in Kent worth £3000 a year, to add to £8000 from the Revesby estate and £3000 more from another one in Derbyshire he inherited from an uncle. He also made wise commercial and industrial investments. In 1781 he was created a baronet, like Hans Sloane; virtually every summer the family migrated to Revesby, where he lived the life of a country gentleman.

In 1797 he was sworn of the Privy Council, on which he was active for a number of years. His voyage in the *Endeavour* was the start of a lifelong interest in Australian affairs, and he took an active interest in the establishment of the first colony at Botany Bay in 1788, a project which he had first urged in 1779, although not as a penal settlement. He declined an invitation to join the administration because he thought he would be more useful in an uncommitted position. Gout was an increasing problem and it became difficult for him to get about. He died in London on June 19, 1820, leaving no legitimate descendants. The Soho Square house, one of London's finest Adam houses, underwent several changes before being demolished in 1937.

Banks constantly acted as unofficial scientific adviser to the government. An important manifestation of this was the part he played in establishing the now world-famous Royal Botanic Gardens at Kew, of which he became unofficial director. His efforts to bring to Kew seeds and plants from every part of the globe were prodigious, and made possible only by his unique international standing. He conceived Kew as not only a centre for maintaining living specimens of as many species as possible, but as offering an advisory service for making practical use of plants. For example, the tea plant was established successfully in Ceylon and India from plants descended from 2000 Chinese originals and 17 000 seeds obtained by the collector Robert Fortune. Banks also tried to establish breadfruit from Tahiti as cheap food for slaves on the West Indies plantations – initially frustrated by the famous mutiny on the *Bounty*, which was carrying the plants. At the king's request he took an active part in arranging the import into Britain of the merino sheep from Spain; sheep of that breed were then closely guarded by the Spanish government.

Banks encouraged, or at least acquiesced with good grace, in the foundation of the exclusive Linnean Society, only second to the Royal Society but devoted to botany and zoology. However he attempted to block the formation of the Geological Society, founded in 1807, and of the Astronomical Society, founded in 1820, seeing these as offering too much of a challenge to the role of the Royal Society. He also opposed access by the general public to the British Museum; in his opinion the function of the museum, of which (as president of the Royal Society) he had been an ex officio trustee for over forty years, was to facilitate the researches of members of the learned societies. Banks used his extensive personal network of correspondents to channel to the museum many gifts of material he received. In his lifetime he presented it with his own ethnographic and zoological collections and in his will his remaining collections and library were to be left as they were until Robert Brown died, then they were to be transferred to the museum; in fact they accompanied Brown when he moved to the museum in 1827. Banks maintained a huge correspondence which is of great interest to historians but his scientific publications do not amount to much.

JEAN-BAPTISTE DE LAMARCK (1744–1829)

In the first half of the nineteenth century the leading exponent of evolution was the French naturalist whos full name was Jean-Baptiste-Pierre-Antoine de Monnet, chevalier de Lamarck. He was born on August 1, 1744, to Marie-Françoise de Fontaines de Chuignolles and Philippe Jacques de Monet de La Marck. His parents were among the semi-impoverished lesser nobility of northern France. They lived in the small Picard village of Bazentin-le-Petit. His father, following family tradition, served as a military officer. It was primarily economic and social considerations that led his parents to select the priesthood as the future career for the youngest of their eleven children. Lamarck, at about the age of eleven, was sent to the Jesuit school at Amiens; he was not much interested in a religious career and much preferred the military life of his father and older brothers. When his father died in 1759, Lamarck left school in search of military glory. He soon rose to the rank of lieutenant and fought with the French army in the Seven Years War. After the war was over, he spent five years at various French forts on the Mediterranean and eastern borders of France. It was during this period

that he began botanising; his military transfers served to acquaint him with the diverse types of French flora.

In 1788 Lamarck left military service because of an injury and after several years of poverty found clerical work in a bank. He subsequently studied medicine for four years during which he became interested in meteorology, chemistry and botany. This last became his main scientific interest and he developed an expert knowledge of the subject with the help of Bernard de Jussieu. In 1778 he published the results of over nine years of work in his elaborate four-volume *Flora française*, arranged as a dichotomous key for the determination of species, which was highly praised and went through several editions. As a result he became a protégé of the influential Buffon; Lamarck was elected to the Académie Royale des Sciences in 1779 and two years later he was appointed royal botanist, charged with visiting foreign botanical gardens and museums, and establishing contacts between their directors and their opposite numbers in Paris. As we have seen, Lamarck took Buffon's son with him on a tour of these institutions. On his return Lamarck wrote voluminous botanical contributions for the *Encyclopédie methodique*, published in 1785. In 1788 he was posted to the Jardin du Roi as keeper of the herbarium, on a paltry salary.

Although Lamarck began his scientific career as a botanist he contributed to other fields of science. Meteorology was one in which he prepared a memoir for the academy which was well received, although it was never published. This led to him taking an interest in chemistry, where he emphasised the fundamental distinction between the organic and inorganic sides of the subject. He was an expert on conchology and formed a fine collection of shells, which eventually he sold to the Jardin des Plantes. This interest led to the study of fossils, and their implications for geology, especially the extent of geological time, which he thought must be almost unlimited.

When the Muséum d'histoire naturelle was created in the reorganisation of 1793 he was appointed to one of two professorships of zoology; the other went to the much younger Geoffroy de Saint-Hilaire, whose profile follows later in this chapter. The museum became an institution where teaching and research occupied an equal place to the care of the collections. It was agreed that Geoffroy should specialise in the vertebrate animals, Lamarck in the invertebrates. Although Lamarck had built up a high contemporary reputation as a botanist in the previous quarter-century, it is the work done after this change of subject for which his name is mainly known today. Almost every thing he did led to the theory of evolution, which he first presented in a lecture at the Academy in 1800. It was about this time that the term biology began to be used instead of natural history. Lamarck was one of the first to adopt it.

In its original form, Lamarck maintained that nature formed the simplest plants and animals directly, through spontaneous generation. Then changing circumstances and needs led to new responses which eventually produced new habits; these habits tended to strengthen certain parts or organs through use. Gradually new organs or parts would be formed as acquired modifications were passed on through reproduction. In 1815 he began publication of his *Histoire naturelle des animaux sans vertèbres*, in the first volume of which he amplified the views on the evolution of animals that he had already stated in 1809 in his *Philosophie zoologique*. Others, such as Buffon and Erasmus Darwin, had already expressed opinions favouring the theory of the evolution of animals, but Lamarck crystallised his ideas in four laws under which evolution takes place. The essential part of his theory is contained in his second law, that 'the production of a new organ in an animal body results from a need which continues to make itself felt'. This has been misinterpreted; and discussion has concentrated on his fourth law, which postulates the inheritance by an individual's progeny of characters acquired during its lifetime. Lamarck's naïve observations on the inheritance of acquired characteristics lasted for over a century, in spite of all the evidence against it.

When Lamarck's views were re-examined after publication of *On the Origin of Species* they were misunderstood and ridiculed. His reputation suffered through the misunderstanding of what he had written, and because few of his critics took the trouble to consult his writings themselves, most of them swallowing unverified the second-hand opinions of others. Lamarck wrote that new needs are satisfied by the development of new structures; the word 'besoin' was mistranslated as meaning 'want', in the sense of desire, and critics who had not read the original ridiculed the suggestion that the giraffe got its long neck because it consciously wanted one, instead of, as Lamarck actually wrote, because it needed it. Furthermore the inheritance of characters acquired during the lifetime of the individual has never been proved, and thus Lamarck's fourth law has been cast aside in the belief that it alone was the essence of Lamarckism. It is a pity that Lamarck's name should be universally known through ideas rejected with scorn by those imperfectly acquainted by them.

By this time Lamarck was fifty years of age. Twice married, he was already the father of six children. In 1777 he began a liaison with Marie Rosalie Delaporte, marrying her, fifteen years and six children later, as she was dying. In 1793 he married Charlotte Victoire Reverdy, by whom he had two children; she died in 1797, after which he married again, once or twice. He was required to live close to his work, in the Maison de Buffon, but he supplemented this by purchasing a small country house at Héricourt-Saint-Samson, about fifty miles north of Paris. His professorial salary was not enough to support a large family and since the Académie des Sciences had been suppressed he lost the pension he received from that source. He was offered the chair of zoology at the university in 1809, but declined because he felt too old to fill the post worthily. Moreover his previous good health had begun to fail, and he was beginning to have trouble with his sight.

In 1818 he suddenly became completely blind, but was able to carry on with his professorial duties with the help of one of his daughters who devoted herself to looking after him during the last decade of his life. When he died in poverty, on December 28, 1829, his family could not afford to pay for his funeral and had to apply to the academy for funds. His belongings, including his books and scientific collections, were sold at public auction; he left five children with no financial provision. Of these one son was deaf and another insane; his two daughters were single and without support. Only one child, Auguste, was financially successful, as a civil engineer; he was also the only one of the offspring to marry and have children. The official eulogy prepared by Cuvier for the Academy condemned Lamarck's speculations and theories in every field as being quite unacceptable. This was true for what he wrote about physics and chemistry, but in biology his work has proved of lasting importance. He made significant contributions in botany, invertebrate zoology and palaeontology, and developed one of the first thoroughgoing theories of evolution.

While he was ignored by his own countrymen, he received some attention in England from the generation before Darwin. But it was really Darwin's theory of evolution which ensured Lamarck's fame. The question of the extent of Lamarck's influence on Darwin is still debated. It was mainly Darwin's enemies and detractors who revived Lamarck for a variety of reasons, ranging from scientific to religion to nationalism (on the part of the French). Towards the end of the nineteenth century a famous controversy arose between the Darwinians and the neo-Lamarckians; the latter used Lamarck's views selectively and often modified them to suit their purposes. Neo-Lamarckism had strong proponents in France, Germany, England, America, and most recently in the Soviet Union. With the wide acceptance of Darwinism as modified by modern genetic theory, much of Lamarckism has died out, although some still apply it to seemingly purposive biological behaviour.

GEORGES CUVIER (1769–1832)

Georges Cuvier, the elder of two scientific brothers, was born in the chief town of the tiny principality of Montbéliard, on August 23, 1769. Montbéliard was geographically French but had been detached from Burgundy in 1397 and placed under the rule of a junior branch of the Grand Duchy of Wurttemburg. During the sixteenth century its inhabitants adopted Luther's doctrines while keeping the French language. Cuvier's father Jean-Georges, who had been an army officer in the service of France, was married late in life to Anne-Clémentine Chatel, a woman twenty years his junior. Their first son was baptised Jean-Léopold-Nicholas-Frédéric, but always known as Georges. The family were not well off since Jean-Georges had already retired when the future naturalist was born, and Anne-Clémentine was in poor health. Very weak at birth, Cuvier remained in delicate health for a long time. During his childhood he enjoyed drawing and gave evidence of a precocious intellectual and emotional development.

Gifted with an astonishing memory he mastered those parts of Buffon's natural history that had appeared.

At the age of twelve he began to make natural history collections and founded a scientific society with some friends, as adolescent prodigies often do. Cuvier's parents intended him to become a Lutheran minister but his schoolteachers thought otherwise. Fortunately the wife of the governor of Montbéliard recommended him to her brother-in-law, the reigning duke, who was seeking able young men to attend the Caroline Academy which he had founded near Stuttgart. The academy was intended primarily for training bureaucrats. The students came from all over central and eastern Europe. The curriculum was broad, the discipline strict, the ethos cosmopolitan. Cuvier entered the academy in 1784, at the age of fifteen. After two years of general studies, during which he learned German, the bourgeois Cuvier was promoted to the rank of chevalier, which allowed him to live with students of noble birth and become known to the duke himself. Thus this young man with bright blue eyes, thick red hair, heavy features and dishevelled clothing began his education as a select member of court. He decided to specialise in administrative, juridical and economic sciences, which included a significant proportion of natural history. With friends he founded a natural history society that awarded decorations to its most active members. In his second year at the academy Cuvier had discovered near Stuttgart some plants that were new to the region. The leading biologist at the academy was the young lecturer in zoology, Karl Friedrich Kielmeyer, who became one of the founders of the German Naturphilosophie. Kielmeyer taught Cuvier the art of dissection and probably some comparative anatomy as well.

When Cuvier completed his studies in 1788 he found there were no vacant positions in the ducal government for a penniless young commoner. Instead he took a position as a private tutor in Normandy, with a noble and affluent Protestant family named d'Héricy. On the journey through France by stagecoach the luxury of Paris astonished him. Revolutionary unrest was beginning, but in Normandy, Cuvier was not affected by the dramatic events that were taking place in the French capital. His duties as tutor were not very demanding. During the autumn and winter the d'Héricy family lived in Caen, where he had access to good libraries and a botanical garden. In the spring and summer they migrated to their chateau in Fiquainville, near the sea and the fishing port of Fécamp, which gave him an opportunity to dissect marine organisms and sea birds.

During his six years in Normandy Cuvier maintained contact with the friends he had left behind at the Caroline Academy through correspondence with Christian Heinrich Pfaff. Since he ran the risk that letters would be inspected by the French police, Cuvier was forced to feign sympathy with revolutionary ideas. After the Revolution, however, he often expressed his disapproval of a regime in which, he said, the populace made the law. He dreaded the populace throughout his life. For the historian of science the Cuvier–Pfaff letters are of great interest as showing how, between the ages of nineteen and twenty-three, Cuvier acquired the basic ideas that he went on to develop later. He began by being generally suspicious of any theories, whether scientific, philosophical or social, telling Pfaff in 1788 'I wish everything that experience shows us to be carefully disassociated from hypotheses . . . science should be based on facts, not systems.' Here he was referring to the German Naturphilosophie, which maintained that, in spite of a near-endless variety of form, all vertebrates are constructed upon one and the same Bauplan, and were to be regarded as manifestations of the same fundamental type. Three years later he explained to his friend that the structure of an animal is, of necessity, in harmony with its mode of life; the form of each organ is related to its function, not to its situation in some master plan. In between Kielmeyer, who had left the Caroline Academy for a short while, returned to it. Pfaff sent details of Kielmeyer's unpublished courses to Cuvier, who suddenly became a convert to the theory known as the chain of being, advocated by Kielmeyer in his lectures, in which organisms are arranged in order of increasing complexity, with *Homo sapiens* at the top. They did not appear all at once but developed gradually, step by step.

Meanwhile the increasingly radical revolution was starting to affect Caen. Cuvier had become a French citizen on the annexation of the territory of Montbéliard in 1793. He then seems to have obtained an administrative position in local government which he used to assist the d'Héricy family to weather the political storm. At the same time he sought recognition in the scientific world by sending a selection of his unpublished work to some of the leading biologists of Paris. One of them was Geoffroy, who as we know had just been appointed professor of zoology at the Muséum d'histoire naturelle. Geoffroy encouraged Cuvier to move to Paris and work with him; and in 1795 Cuvier did so. Geoffroy was extremely friendly to him upon his arrival in Paris and for the first year they lived and worked together. In a joint paper on orang-utans they audaciously proposed the idea of the origin of species from a single type, following Buffon. Shortly after his arrival Cuvier produced an important paper on the classification of invertebrates. Previously naturalists had divided these into two classes, insects and worms, but now, after the research he had done in Normandy, he proposed a much better one, consisting of the molluscs (e.g. clams and cuttlefish), the articulates (e.g. insects and worms) and the radiates (e.g. jellyfish and starfish). There was, he maintained, a 'ground plan' or 'base plan' for each branch of the animal kingdom, and he believed there were four such branches, with no connection between them. This was at once accepted by Lamarck, the other professor of zoology, who announced to his course that he was going to follow to a very great extent the classification devised by 'the learned naturalist Cuvier'. He also agreed with Cuvier about the chain of being at this time, although later he changed his mind.

In Paris, where there was a shortage of zoologists, it was not so surprising that soon after his arrival he should be appointed assistant professor of animal anatomy at the Muséum d'histoire naturelle. Because of this he was provided with quarters in the Jardin near the menagerie, in which he lived until his death. The appointment was the consequence not only of the importance of his scientific work but also of his ability as a teacher. After only a few minutes of preparation he was able to deliver a` logically constructed lecture in a confident manner; without stopping his delivery he illustrated his ideas by means of quick blackboard drawings that were as clear as they were accurate. He captured the popular and scientific imagination of his day by reconstructing the form of large extinct vertebrates from fragmentary fossilised remains. At the end of 1795 Cuvier, at the age of twenty-six, became one of the original members of the First Class of the new Institut de France, which replaced the former Académie Royale des Sciences (the classes were not ranks; they indicated the departments in which the candidate had been elected). Social contacts were as important as scientific merit in such elections.

Cuvier now began to distance himself from Geoffroy, who had been so helpful to him earlier. In 1799 they responded differently to Bonaparte's call for scientists to accompany him on his unsuccessful Egyptian expedition. Geoffroy went to Egypt and had reason to regret it; while Cuvier stayed in Paris and furthered his career. Later he was conspicuously unhelpful to Geoffroy in his attempts to get a seat in the Institut.

In France, before the Revolution, education had been dominated by the church. The Revolution had abolished the colleges and universities and confiscated the property of the teaching orders on which the provision of most secondary education depended. The Directorate did not reverse these acts but introduced a new system of Ecoles centrales, one for each department, which emphasised scientific education. These soon turned into lycées where the traditional curriculum was reinstated. Although Cuvier disapproved of the educational policy of the regime in 1802 he accepted the well-paid post of inspecteur-général des études, in which capacity he was sent to the south of France to organise new lycées at Bordeaux and Marseilles, to inspect other schools in Provence and to examine candidates for places and positions in the new institutions. Disliking the work involved he resigned after a year and assumed the less remunerative but more influential duties of permanent secretary of the First Class of the Institute. As such he became the official spokesman for the natural sciences

The following year the Consulate was replaced by the Empire, essentially a military dictatorship. Although Cuvier strongly disagreed with the policies of Bonaparte he accepted an appointment as inspector-general and member of the council of the Université imperiale. This umbrella organisation had been set up to administer not only the three reformed universities in France itself but also to supervise the existing universities of the countries of the rapidly expanding French Empire outside France. Cuvier was able to use the powers of his new position to reform curricula and to make appointments to senior posts. He also played an important role in the organisation of the new Sorbonne, drawing on what he found out about universities in other countries. He took advantage of his great influence in higher education by trying to develop the teaching of religion, modern languages and the natural sciences. Cuvier renounced the audacious ideas about evolution he held earlier, especially after the return to religious beliefs marked by Napoleon's coronation by the Pope in 1804.

After the collapse of the Empire in 1815, the return of the Bourbon dynasty did not bring back the *ancien régime*. Instead the new system of constitutional government imposed by the victorious allies created a period of political instability. Although unable, as a Protestant, to hold ministerial office, Cuvier's membership of the Conseil d'Etat gave him great power and influence, which he used to obtain positions for friends and relatives. For himself, contemporaries said, he accumulated more well-paid government posts than any single man had a right to. Cuvier also became a member of the literary Académie Française in 1818, was made a baron in 1819, a Grand Officer of the Légion d'honneur in 1824 and a peer of France in 1831.

In 1804 Cuvier had married a widow, Anne-Marie née Coquet de Trayzaile, a devout Protestant who was described as 'kind, outspoken and energetic'. Already the mother of four children, she bore him four more. It is his stepdaughter Sophie Duvaucel rather than his wife who features in memoirs of Cuvier as his faithful assistant and hostess. Neither of Cuvier's sons had survived childhood; only one of his daughters, Clémentine, did so, and she was to die in 1827 a few weeks before her marriage could take place. She was particularly close to him (it is conjectured that he used her to make charitable donations on his behalf) and following the death of this last of his four children he plunged himself into his work as never before.

Although remarkably vain Cuvier was short in stature and not particularly good-looking; he gave himself a commanding appearance by dressing up. In youth he had been slim but he put on weight during the Empire and became enormously fat after the Restoration. He had to walk slowly and not bend over for fear of apoplexy. His health and his appetite remained excellent, however; his favourite Montbéliard chitterling sausages were never missing from his table. Cuvier was always impatient, easily irritated, pliant towards those he regarded as his superiors, authoritarian towards those he regarded as inferiors. He was a very secretive man and it is not certain that his writings reflected his real opinions. Nevertheless he was kind to the aspiring young, assisting and advising them. Cuvier had several large sources of income, any one of which would have enabled him to live comfortably. He had a carriage and servants, visited the Paris salons, and himself received at home in the great hall of his library, furnished with busts of the famous. His appetite for scientific and administrative work was immense and as he grew older it became greater. This was possible because of his extraordinary memory. He knew the contents of almost everything in his huge library from which he could retrieve information in seconds, but this seems to have made it difficult for him to synthesise. Cuvier sought to do the most possible in the shortest space of time, and therefore he rarely sought perfection in form or thought. One evening in May 1832 Cuvier began to feel ill. He grew progressively weaker over the next few days and died on May 13.

Cuvier wrote an autobiography, of which the original version has been lost. It was so edited by his widow and one of his colleagues that it gives an inadequate picture of some of the critical periods in his career, as do some of the memoirs published after his death. We are on firmer ground with his scientific work. In zoology this resulted from his dominant position at the Muséum d'histoire naturelle, which at the time was the world's largest establishment devoted to scientific research. He rearranged the collections according to his own theories and organised expeditions to bring back fresh material, but travelled little for this purpose himself. In order to gain time he surrounded himself with collaborators who lacked his intellectual vigour and who would not dare to criticise him; thus numerous works published under his name are not all of high quality. However his ichthyological work proved to be of lasting value, and he greatly advanced palaeontology. Although he studied the fossil record carefully his erroneous theories brought him into conflict with his former friends Lamarck and Geoffroy. Cuvier not only fought them in public, he also used his political power: Geoffroy, and probably also Lamarck, became the object of investigation at a time when orthodox religious beliefs were obligatory for all civil servants.

In his youth, as we have seen, Cuvier had first disputed and then accepted the theory of the chain of being, but later he came round in favour of the fixity of species and separate creation. He decided that the theory of the variability of species was contrary to moral law, to the Bible and to the progress of natural science itself. In his last lecture, six days before his death, he pronounced anathema on useless scientific theories. By this he meant not only pseudo-sciences like mesmerism and phrenology but also the theories of evolution held by respectable biologists such as Lamarck and his disciples. He rendered solemn homage to divine intelligence before an audience overcome with emotion.

ALEXANDER VON HUMBOLDT (1769–1859)

The father of the Humboldt brothers was Alexander Georg von Humboldt, a major in the army of Frederick the Great, who was wounded in 1761 during the Silesian war. He then served at the Court in Berlin and was appointed chamberlain to the Crown Princess. He married Marie Elisabeth von Holwede, a widow, née Coulomb. She was descended from a noble Burgundian family who, being Huguenots, had settled in Berlin after the passing of the Edict of Nantes. She brought to her second marriage a tenanted farm in the Neumark, an estate called Tegel, in picturesque country near Berlin, and a house in Berlin itself. Wilhelm was born on June 22, 1767 in Potsdam, Alexander on September 14, 1769 in Berlin. There was also a daughter who died in infancy, and a son by their mother's first marriage. Their father died when Wilhelm was twelve and Alexander ten, after which their most capable mother supervised their upbringing, planned from early childhood. As a boy Alexander was much less robust than his brother; both suffered from episodes of depression, as did their mother.

Until they went to university, when their paths separated, they were educated at home by carefully selected tutors. However they also participated in the flourishing Jewish society of Berlin, led by Moses Mendelssohn and known as the Berlin Enlightenment. A small group of highly intelligent Jewish families who had made fortunes in trade and industry showed an active interest in literature, the arts and the sciences and would discuss them in their salons, particularly those of Henriettte Herz and Rahel Levin. Alexander had a crush on Henriette, considered the cleverest and most beautiful young woman in Berlin.

The education of the Humboldt brothers was mainly directed towards preparing them for service to the Prussian state. They began their university studies at Frankfurt-am-Oder, where there was a short-lived institution which catered for Prussians of noble blood who were interested in duelling rather than anything academic. After a year Wilhelm transferred to the Hanoverian University of Göttingen to study law while Alexander stayed on at Frankfurt studying public finance and administration. Already, however, Alexander was developing an interest in botany and the natural sciences generally, which could not be satisfied at Frankfurt, so he returned to Berlin and then after a while joined his elder brother at Göttingen, which was celebrated for mathematics, physical science and medicine. During the long vacation between the two semesters he made the first of the journeys for which he became famous, in the company of the naturalist Georg Forster,

who with his father Johann Reinhold Forster had circumnavigated the globe on Cook's voyage of 1772, and whose account of it in *A Voyage Around the World* (1777) is considered one of the best pieces of eighteenth-century ethnographic reporting. Alexander had felt the urge to travel to distant lands since childhood; he began with a tour of northern Europe with the younger Forster. They travelled through the Low Countries to England, then returned to Germany via Paris. 'We could not be travelling at a more fortunate time,' he wrote to his parents, referring to the first historic stage of the French Revolution. The next stage in Alexander's education was attending a commercial college in Hamburg, where he learned the basics of political economy. The students at the academy came from various parts of Europe; he took the opportunity to add some more languages to those he already spoke.

Having become interested in mineralogy and geology he spent a further year studying at the famous school of mines in Freiberg, where the charismatic Abraham Gottlob Werner (1749–1817) was attracting enthusiastic students from all over Europe. He then secured a position as mining supervisor in Franconia in the Administrative Department of Mines and Smelting Works. It was never his intention to make a career of this but rather to gain practical experience. He distinguished himself by introducing important safeguards for the health and safety of mineworkers and rose to the rank of supervisor-in-chief, only one step below the rank of minister. He was also elected associate of the Imperial Academy of Sciences at Breslau for his scientific work, not only in geology but also in natural history, as displayed by his *Florae fribergensis specimen*. At the end of five years he made a long journey through northern Italy and Switzerland, partly to widen his geological experience but also to meet scholars and other interesting people on the way, including Volta at his villa on Lake Como and Saussure in Berne. He completed a paper on the stimulus of muscles. At this time there was a major controversy about the creation of the earth, between the Neptunists, led by Werner, who maintained that the Earth had come into being through sedimentation in the oceans, and the Plutonians or Vulcanists, led by the Scottish geologist James Hutton (1726–1797), who maintained that the main rock layers of the Earth were of volcanic origin. The critical issue in this debate was the origin of basalt: according to the Neptunists this too was sedimentary, whereas the Vulcanists maintained, rightly, that it was of igneous origin.

When their mother died after a long and painful illness in 1796 she left both her sons sufficiently well off that they had no further need to earn a living. Having like his brother Wilhelm left the civil service, Alexander could now realise his childhood ambition to travel to distant lands. He went with his brother to Weimar to consult Goethe and Schiller, and then to Paris to seek the advice of members of the Académie des Sciences. An invitation from the eccentric Lord Bristol to join him on an expedition to Egypt fell through when it became known that Napoleon was planning a military expedition to the same area. A French plan for a voyage to Australia and possibly around the world was postponed, and the self-effacing veteran Louis de Bougainville replaced by the less experienced Nicolas Baudin, whose entourage included a team of civilians briefed by Cuvier. Humboldt planned a visit to Algiers; that also fell through. Finally Humboldt, by then twenty-nine years of age, decided on Spanish America, and he chose to take with him as his assistant and companion the twenty-five year-old French botanist Aimé Goujard Bonpland.

They set off at the end of 1798 for Madrid to arrange for the necessary permits and letters of credit. Humboldt, who had already become fluent in Spanish, was able to persuade King Carlos IV to grant all the facilities they could wish for. As a result they were very well treated throughout the Spanish possessions, although they were barred from entering Brazil by the Portuguese colonial government because of Humboldt's radical political views. They sailed out of La Coruña in one of the mail ships which regularly crossed the Atlantic to the Spanish possessions in South America. The outward journey took six weeks, including a week in the Canaries, where Humboldt climbed to the top of the huge volcano which dominates Tenerife. He climbed several other high mountains on his journey, mainly to see how the vegetation varied with altitude, and to understand their geological structure. They were all of volcanic origin; one of them, Chimborazo, was then thought to be the highest mountain in the world, but he did not quite reach the summit. Humboldt's journey really began in Venezuela, where he spent fifteen months. The most interesting part was travelling by river, 'living like Indians'. One of the tasks he set himself was to discover whether there was any connection between the Rio Negro and the Orinoco; they decided that there was (in fact there is not). On the return journey to the coast they both became seriously ill, Bonpland almost died.

From Venezuela they crossed the Caribbean to Cuba, where Humboldt stayed three months in Havana. His original plan had been to proceed northward to the Great Lakes and the south again to Louisiana and then Mexico. Since Baudin planned to travel by sea and reach Chile and Peru via Cape Horn, Humboldt decided to cross the isthmus of Panama and meet the French on the Pacific coast. In the end, however, the French expedition never got very far and was abandoned, for financial reasons, and as a result Humboldt decided that instead of travelling down the coast by sea, he would make the long and arduous journey overland, up into the Andes and through Colombia and Ecuador to Peru. Although well treated in Lima, as everywhere they went, they were disappointed by the lack of objects of scientific interest, and left after two months. He then sailed up the coast, reaching Acapulco towards the end of March 1803. Finally he spent a few weeks in Mexico, which he liked so much that he thought of going to live there.

Originally Humboldt had intended to return to Europe via the Philippines and Egypt but decided to keep this for a separate journey to South Asia and India; in fact this was never carried out. Instead he went from Vera Cruz up into the United States via Havana, reaching Philadelphia on May 20, 1804. He was given a warm welcome by President Jefferson, who shared many of Humboldt's interests, and invited him to stay in Monticello for several weeks. Later in life Humboldt always received American visitors with particular courtesy. Finally he sailed home from Philadelphia to Bordeaux. The entire journey had taken well over five years. Although its main purpose had been scientific, there was much else that Humboldt wished to write about in the account of his journey which he started to work on soon after his return. Over a period of five years he had collected an enormous amount of data. One of his most famous illustrations was a foreshortened and bisected view of Chimborazo, half of which was a realistic representation of the forested flanks and snow-capped peak, the other half a chart listing every plant collected on the volcano, with its precise location noted. He was the first to create topographic maps of floral habitats in this way. Humboldt also invented an entirely new genre of travel literature by including objective descriptions of geology and natural history, details of his daily experiences, depictions of the customs and institutions of the indigenous peoples, and subjective impressions, judgements and emotional reactions – all of which he wove into a seamless narrative with polymathic skill, aesthetic sensitivity and literary craftsmanship. Humboldt's methodology, based on the precise and accurate recording of his observations and measurements, and the quality of his written account, set the standard for all future scientific travellers.

While he had been away France had changed from being a republic to a military dictatorship, the centre of a rapidly expanding empire. Although Prussia was at war with France, Humboldt settled down in Paris, living at the Ecole polytechnique and working at the Académie des Sciences. The lectures he gave about his travels were much appreciated. The scientific academicians gave him good advice and François Arago, the permanent secretary, became a close friend. He assembled a team of assistants and illustrators; they produced seventeen folio, nine quarto and seven octavo volumes, written in French, although some sections were translated into German. Sixteen volumes deal with botany: 8000 species of plants are described, of which over half were previously unknown. However, the American journey had seriously diminished his wealth, and there was also the cost of producing his magnum opus, which was not a commercial success. In addition the brothers had lost a lot of money invested in Polish property. When Humboldt was trying to raise funds for his intended expedition to central Asia and northern India he made a visit to England hoping to receive the support of the East India Company but without success, although he was lionised by the Royal Society. Eventually the plan was abandoned. Instead Humboldt made a journey to Russia in 1829, but this was of little scientific interest. He also made plans for a pan-American scientific institute in Mexico City, but although the Mexicans were enthusiastic he concluded that it would be too difficult to assemble a group of important scientists which would justify his emigration to that country.

Two years before Humboldt had reluctantly left Paris and returned to Berlin, where his arrival was greeted with great popular enthusiasm. The King appointed him to the undemanding but well-paid office of royal chamberlain, granting him leave to spend four months of each year in Paris until 1848. He was also made a member of the Prussian Academy with the stipend of a university professor but none of the duties. When called upon he undertook diplomatic missions on behalf of the Prussian government but he refused to accept an official position as ambassador, fearing it might jeopardise his good relationships with foreign scientists. Humboldt's long-standing friendship with the Crown Prince continued when he succeeded to the throne as Frederick William IV in 1840. The new king, unlike his predecessor, was a dilettante in the arts but little interested in the sciences. He was too indecisive to be a successful ruler, leaving too much to his reactionary ministers. Humboldt's duties at court became, with the years, increasingly burdensome; the king expected him to dine with him every day and was annoyed if he was absent from a meal without good excuse.

Early in 1857 Humboldt suffered a minor stroke, but made a good recovery. In October 1858 he was severely ill with influenza from which he never really recovered, being then in his ninetieth year. During the last year of his life 1859, he was often confined to bed, and died on May 6 that year. His elder brother Wilhelm had died in 1835, after a highly successful career in the service of the Prussian state; his greatest achievement was when, as Minister of Instruction under Frederick William III, he planned the model of a modern university for Berlin. The younger Humboldt maintained an enormous correspondence, only some of which has been published. He was a sociable person who liked the sound of his own voice but according to Arago he had a very sharp tongue. He lived in the age of the salon and was a regular visitor to those held in Berlin and Paris. He was also remarkably generous to those he considered merited it, so much so that although his mother left him plenty of money later in his life he needed to borrow from friends.

His magnum opus, the account of his American journey, is not the most important and influential of his books. These are the *Kosmos* and the *Ansichten der Natur*, in which he described his understanding of the natural world, a view which he shared with Goethe. Humboldt made no important scientific discoveries, but Charles Darwin said 'I always admired Humboldt, but now I worship him.' Humboldt never married; instead he maintained a long-lasting relationship with a Lieutenant Reinhard von Haeften, four years his junior. When he died he left the larger part of his fortune to his valet. Over the past fifty years the question of his sexual orientation has been much debated. His political opinions might be described as radical and humanitarian; in his strong opposition to slavery, based on what he had seen of it in Spanish America, he was well ahead of his time.

4 From Geoffroy to Hooker

ETIENNE GEOFFROY DE SAINT-HILAIRE (1772–1844)
Geoffroy's father, a procurator at the tribunal of Etampes, a small town near Paris, had little money and fourteen children to support. Etienne, the youngest, was born on April 15, 1772 and was given the surname of Saint-Hilaire which he later joined to his patronymic, although he is generally known as Geoffroy. His career was furthered by scientific priests who were captivated by his lively intelligence, his unusual imagination and physical charm. Thanks to them he was made a canon at the age of fifteen with prospects for a splendid career in the church. However he was already becoming interested in natural history, influenced by the botanist Antoine de Jussieu at the Collège de Navarre in Paris, where Geoffroy was on a scholarship.

With the outbreak of the Revolution ecclesiastical careers were in jeopardy. Geoffroy's father decided he should study law although his own preference was towards medicine. By 1792 he was a pensionnaire libre at the Collège du Cardinal Lemoine in Paris. There he won the affection of the most illustrious member, Abbé René Just Hauy (1743–1842), one of the founders of the science of crystallography. In Hauy's simply furnished quarters he met all the famous scientists of the period, among them his great friend the venerable L. J. M. Daubenton (1716–1800), under whom he studied mineralogy at the Collége de France. While retaining a deep attachment to the priests who supported his early career, Geoffroy embraced revolutionary ideas with characteristic enthusiasm. He frequented the clubs and committees and adopted a philosophical deism and a generous humanitarianism that he preserved for the rest of his life. When the Terror began, in August 1792, Hauy was imprisoned because he was a priest; by actions as courageous as they were romantic, Geoffroy tried to free him, and finally Hauy was liberated. In recognition of Geoffroy's efforts Daubenton arranged for Geoffroy to be appointed demonstrator at the Jardin des Plantes, replacing the count de Lacepède (1756–1825), who as an aristocrat had been forced to flee from Paris.

In June 1793 the Jardin des Plantes became part of the Muséum d'histoire naturelle; and almost at once, owing to the influence of Daubenton, Geoffroy was appointed to one of the professorships of zoology, when barely twenty-one years old, while Lamarck, as we know, was appointed to the other. Each of them eagerly explored his new field. At this period the Abbé Tessier, Geoffroy's former patron, recommended to him a poor young man living in Normandy who was making excellent drawings of careful dissections of fish and invertebrates; this was Cuvier. Geoffroy invited the young man to work with him in Paris and was impressed by his ability. Until then Geoffroy had been hostile to the notion of a great chain of being but now changed his mind, probably under the influence of his collaborator. To support the idea intermediate forms were needed; for example they placed tarsiers between apes and bats in the chain of being. The unity of anatomical structure of vertebrates as a whole had been suggested as early as 1753 by Buffon, but in Geoffroy's time it was still controversial. To Geoffroy the unity of the vertebrate plan implied a sort of kinship among the vertebrates, the more complex having descended from the simpler: thus the mammals are descendants of the fishes. Cuvier would have none of this since he believed, for example, that the fish's fin bore no relation to the mammal's paw, each having been created separately by God. Fifteen years of work beginning in 1806 enabled Geoffroy to establish two fundamental principles of comparative anatomy: first the principle of anatomical connections, which allows an organ to be traced from species to species despite its transmutations, and second the principle of balance, which manifests itself in a reduction in the size of organ when a neighbouring organ hypertrophies.

In 1796 Geoffroy, no doubt under Lamarck's influence, began to investigate modification of species due to the environment. His work was interrupted by Bonaparte, who was organising his ill-fated Egyptian campaign. In response to his request for scientists to join the expedition Geoffroy accepted enthusiastically, and from 1798 to 1801, in the midst of adventures in which he often risked his life, he proceeded up the Nile as far as Aswan, making scientific observations and collecting material to take back to Paris. The English allies of the Egyptians were victorious but Geoffroy succeeded in rescuing his collections from the English. On his return to Paris he found that Cuvier, who had not responded to Bonaparte's call, had become France's leading naturalist in the eyes of the French public

Despite his predilection for sweeping speculations Geoffroy devoted himself from 1802 to 1806 to descriptive zoology and classification. He returned to the research on marsupials which he had begun in 1796. He also composed a large catalogue of the mammal collection of the museum, the printing of which he stopped suddenly in 1803, perhaps as a result of differences with Cuvier; the work remained unfinished. Four years later Cuvier withdrew his opposition and Geoffroy was elected to the Académie des Sciences.

The study and publication of the rich material Geoffroy had gathered in Egypt proceeded at a leisurely pace, since it formed part of the sumptuous *Description de l'Égypte par la Commission de sciences,* published in 1808–1824. In Egyptian tombs Geoffroy had found mummified animals more than 3000 years old, which were identical to existing species. Cuvier saw in this fact the proof of the fixity of species but Lamarck thought this was much too short a time to have any bearing on the matter. Although Geoffroy does not seem to have accepted Lamarck's views on the duration of geological epochs, he continued to believe that species were transmuted in the course of time, the simpler ones engendering the more complex, and in the following years he strove to prove this.

To obtain evidence for transmutation Geoffroy turned to comparative embryology. Kielmeyer, Cuvier's teacher at the Caroline Academy, had discovered that vertebrates, in the course of their embryonic development, go through stages recalling their supposed ancestors: for example, the human embryo develops preliminary forms of the bronchial fissures typical of fish.

According to Cuvier, Geoffroy sought, by intervening in the development of the chicken embryo, to maintain the fish stage or to develop the transition to the mammalian stage. He failed, but his attempts make him the founder of experimental embryology. In the scientific language of the early nineteenth century, the word 'evolution' came to mean the sum of the transformations undergone by an embryo. The term was then promulgated internationally by one of Geoffroy's disciples, Frédéric Gerard, in the *Dictionnaire universel d'histoire naturelle.* In 1831 Geoffroy adopted this new meaning of evolution, which implies the transmutation of species. He attributed the variations between species to physicochemical changes in the environment in the course of geological time, changes that had influenced not the adults, as Lamarck had thought, but the embryos.

Geoffroy also thought, as did some botanists of the preceding century, that the appearance of monsters might provide an explanation for sudden transformations of species. From 1802 to 1840 he devoted more than fifty reports to descriptive teratology; moreover, his *Essai de classification des monstres* of 1821 marks the debut of scientific teratology. In that year Cuvier introduced into the second edition of his *Recherches sur les ossaments fossils*, a study, done a little too hastily, of the remains of a crocodile discovered near the city of Caen. Geoffroy, who had studied the living Egyptian crocodile extensively, announced that the Caen animal was in reality very different from the crocodile. He named it *Teleosaurus*. It presented, he stated, characteristics intermediate between those of saurians and mammals, suggesting a new way for Geoffroy to demonstrate the transition from reptiles to mammals. Although modern studies have somewhat modified Geoffroy's interpretation, it does nevertheless mark the starting point of evolutionary palaeontology.

The greatest problem for evolutionists was how the transition from the invertebrates to the vertebrates was accomplished. Carried away by his imagination Geoffroy made the improbable suggestion that the carapace of insects corresponds to the vertebrae. In 1829 two young researchers proposed an ingenious interpretation to account for the transition from the cephalopods to fish, and Geoffroy used this as an opportunity to attack Cuvier at the Paris Academy. This initiated the famous controversy that so excited Goethe which began on February 22, 1830 and continued until the end of Cuvier's life. Geoffroy was worsted in each of their encounters and withdrew, so overpowered by his opponent that he no longer knew where he was. Over and over again Cuvier, with his prodigious memory, had every fact at the tip of his tongue, and was able to rebut the speculative character of Geoffroy's arguments, which were not supported by sufficient evidence.

Geoffroy's last contribution to biology was his investigation of the fossil mammals discovered in the Perrier bed (old Pleistocene) in the Massif Central. He established that all of them had disappeared from nature and concluded that certain of them constituted intermediate links in the chain of being. After this his writings became increasingly theoretical and vague, infused with the kind of cosmic poetry born of the great, if somewhat mystical, idea of the fundamental unity of the Universe. The Academy reacted by publishing just the titles of his communications. In 1840 he became blind as the result of cataracts. His mental powers began to fail, and he died in 1844, at the age of sixty. It is regrettable that in palaeontology Geoffroy never followed up his speculations with detailed and illustrated memoirs in support of his theories, which after his death his son Isidore edited to form the first scientific treatise on teratology.

Geoffroy, like his friend Lamarck, lived too early to be completely understood. He revitalised comparative anatomy in France and he created scientific teratology, experimental embryology and the concept of palaeontological evolution. Moreover, in judging him only on the basis of his publications, there is a risk of underestimating his influence, for his contemporaries agree that he was bolder and clearer in his speech than in his writings. Much loved by his students, he put forth his ideas for forty-seven years in his courses at the museum and for thirty-two years at the Sorbonne. In his prime the leading liberal thinkers of Paris frequented his home, and the news of his death in 1844 was received with great sorrow by the many who were inspired by his enthusiasm and his liberal ideas. In France today, and perhaps even more in English-speaking countries, Geoffroy's work is often ranked below that of Cuvier, his implacable adversary, yet even Cuvier recognised that Geoffroy possessed a great talent for description and classification.

ROBERT BROWN (1773–1858)

The man described by Alexander von Humboldt as 'botanicorum facile princeps' was born on December 21, 1773 in the historic Scottish town of Montrose. He was the second and only surviving son of the Reverend James Brown, a Scottish Episcopalian, and Helen, daughter of the Reverend Robert Taylor. The father was a man of strong independent views; the son inherited these and retained his intellectual honesty and sturdiness of character, but lost his uncompromising religious faith. Robert was educated at Montrose Grammar School and Marischal College, Aberdeen, and then studied medicine at Edinburgh University. The whole Brown family had moved to the Scottish capital, where James Brown led a tiny religious sect

until his death in 1791. Although he failed to graduate, Robert attended lectures on natural history, made field trips into the highlands, and discovered several new species of Scottish flora. Brown then enlisted as an ensign in the Fife Regiment of Fencibles in 1795 with the duties of assistant surgeon, and accompanied the regiment to Ulster.

In October 1798, apparently while in London on a recruiting mission, Brown was introduced to Sir Joseph Banks by the botanist Jose Correa da Serra, then in exile from Portugal, who referred to Brown as 'a Scotchman, fit to pursue an object with constant and cold mind'. Brown impressed Banks and Drysander, his erudite botanist–librarian, with his zeal and ability. Accordingly, when in December 1800 plans had matured at the Admiralty for a voyage, commanded by Matthew Flinders, to survey the southern and northern coasts of Australia, Banks, whose opinion carried much weight at the Admiralty, offered Brown a recommendation for the post of naturalist aboard the naval vessel *Investigator*, at the very substantial salary of £420.

Brown immediately accepted this attractive offer. He came to London and, until the sailing of the *Investigator* in July 1801, spent his time studying the specimens, illustrations and literature about Australian plants available at Banks' house. Flinders was gathering his crew together; for Brown the most important other member of the ship's company was the botanical draughtsman Ferdinand Bauer, selected by Banks. The outward voyage was uneventful although the *Investigator*, an old and leaky vessel, was not ideal for the task ahead; nothing better was available because of the needs of the Royal Navy in the war with France. They called briefly at Madeira and other islands en route, and for rather longer at the Cape of Good Hope, where Brown was able to study the proteas, found only in the southern hemisphere. Eventually they reached the south-west corner of Western Australia at King George Sound near the present town of Albany. There Brown was fascinated by the astonishing richness of the flora, its plants in their diversity and strangeness far exceeding anything previously seen. Brown collected some 500 species, almost all of them new to science. From Lucky Bay, which yielded 100 more species, the *Investigator* sailed eastward along the southern coast of the continent, passed through the Bass Strait, between Victoria and Tasmania, and turned northward to Port Jackson, close to Sydney, which was reached in May 1802.

Most of the seeds and other material that had been collected was sent off to London by the next available ship. After eight weeks the expedition continued northward to Cape York and into the Gulf of Carpentaria. Unfortunately *Investigator* had been damp, leaky and unsound from the start, and Flinders dared not continue his survey. In addition there was much sickness on board: the captain himself was suffering from scurvy. Flinders sailed to Timor for provisions, then returned to Port Jackson, arriving there in June 1803, where it was doubtful whether the ship could be repaired. He then started for home in another ship carrying valuable plant and other material that had been collected. Unfortunately before they had gone far the ship struck a reef in the Torres Straits and all the material was lost, while everyone on board returned to Port Jackson with difficulty. On his journey home Flinders' ship was captured by the French, who were at war with England. He was then detained on the island of Mauritius and despite Banks' efforts to secure his release he did not arrive back in England until years later.

Meanwhile Brown had gone to Tasmania, where he found much interesting new material. When the *Investigator* had been repaired, as far as possible, they sailed non-stop back to England across the Pacific, reaching home in October 1805, just before the news of Nelson's victory over the French at Trafalgar. By this time Brown had been away over four years, and the *Investigator* was only fit for the scrapyard. There was some delay over Customs clearance for their collection of specimens of nearly 4000 species of plants, as well as drawings and zoological specimens including a live wombat.

During the next five years Brown described nearly 2200 species, over 1700 of which were new, and had selected about 2800 of them for the British

Museum. At the same time Brown was serving as clerk, librarian and house-keeper to the Linnean Society. The period from 1806 to 1820 was that of Brown's greatest creative endeavour, when he worked under Banks' fatherly eye. Also his planned two-volume *Prodromus Florae Novae Hollandiae* (in somewhat faulty Latin) never got beyond the latter half of the first volume. There is no doubt that Brown was serious about publishing the whole work but was discouraged when he was informed of some quite minor errors in the text. He recalled the unsold copies which had been printed and corrected some of them by hand, but not many were sold. The full account of the botanical discoveries made on the Flinders expedition never appeared. Hardly any botanical works of this type were commercial successes, although there was a market for lavishly illustrated books on botany.

When Banks' resident librarian Drysander died, Brown was appointed his successor, while continuing to work part-time for the Linnean Society; he was no longer employed by the Admiralty. Brown dined regularly at Banks' table, where he met other scientists and their patrons. He was elected to the Royal Society in 1811 and to the dining club associated with the Linnean. When peace was declared between England and France Brown was one of many English scientists who went to Paris to contact their French counterparts and invite them to visit London. His international reputation was constantly growing. He made a tour of France and Italy in 1824, and in 1828/9 made an extensive tour of France and Germany, visiting sites of special botanical interest and meeting the famous continental botanists. In 1819 the chair in botany at Edinburgh University became vacant and efforts were made to persuade Brown to apply. He was sorely tempted but Banks, then in poor health, did not wish to lose his services. As a result Brown stayed as he was, until on his patron's death the next year he found that he had been bequeathed an annuity of £200 and, subject to the life interest of his widow, the life tenancy of Banks' house, with the use of its library and collections. On Brown's own death these were to pass to the British Museum but, as we shall see, Brown arranged the transfer in 1827, on condition that he became the curator of botany in the Museum. All this suited Brown well enough except that the annuity was not paid in full, and a condition was imposed that he would lose it if he took other employment. When the Linnean Society took a sublease of part of the property he was left with the accommodation he needed for his work and the back of the house to live in.

In 1822 Brown resigned from his post of librarian of the Linnean Society and turned down the offer of the position of honorary secretary. Later he became president. He was now fifty-one years of age, frequently unwell; he remained a bachelor to the end of his life. His international reputation was still growing. At home he was elected an honorary member of the Cambridge Philosophical Society and of the Royal Society of Edinburgh. On the occasion of a meeting in Oxford of the British Association for the Advancement of Science, the university conferred an honorary degree on Brown at the same Encaenia as John Dalton and Michael Faraday. Brown was also an honoured guest at meetings of the relevant societies on the continent. He travelled throughout most of Europe, despite the rough conditions before the railway network was fully established. He was elected a foreign associate of the Paris Academy of Sciences, and received similar honours from other scientific academies. At home he was awarded the prestigious Copley Medal of the Royal Society for his work on fertilisation and embryology, and received an honorary degree from Edinburgh University and the freedom of the cities of Edinburgh and Glasgow.

By this time Brown entered into negotiations with the trustees of the British Museum over the future of the Banksian herbarium and library in Soho Square. At this time the museum was being widely criticised for its lack of care of the collections and its weakness in scientific staff. This was one of the reasons why there was a proposal to found a new non-sectarian university in London, and in 1827 Brown was asked if he would accept the chair of botany. He declined but seems to have used the offer as a way to persuade the trustees of the British Museum to appoint him as Under-Librarian (essentially Keeper) of the botanical collections, which would be augmented by those in Soho Square. This was agreed, and he now became Robert Brown of the British Museum. In 1833 he gave evidence to a select committee of the House of Commons into the affairs of the museum which was under increasing attack, accused of malpractice of all kinds, general inertia and inefficiency, nepotism, and neglect and weakness of science. He asked for an assistant, and a study where he could work in private, but especially for additional funds to build up the herbarium. He repeated his request for an exhibition gallery, which was agreed.

When living in London Darwin used to make a point of visiting Brown every Sunday morning: 'He seemed to me chiefly remarkable for the minuteness of his observations and their perfect accuracy. He never propounded to me any large scientific views in botany. His knowledge was extraordinarily great, and much died with him, owing to his excessive fear of never making a mistake. He poured out his knowledge to me in the most unreserved manner, yet was strangely jealous on some points.' Brown was now particularly interested in fossils and palaeobotany. We do not know his views on the origin of species, although it is clear that he did not believe in their fixity.

It was in 1827 that Brown was observing, through his microscope, the random motion of pollen grains and other tiny particles in water, now known as Brownian movement. In fact the phenomenon, which is due to the impact of molecules on each other, was first reported by Ingen-Housz, buried in a paper on another topic, almost forty years before, but Brown was the first to investigate it thoroughly. Brown often found his discoveries were anticipated by other botanists.

Scientists at this period often became embroiled in priority disputes: Brown, a man of high principle, was often caught up in these. For example there was one over the discovery of the chemical constitution of water. To him, and to many others, it was quite clear that Cavendish had priority. Arago, then permanent secretary of the Académie des Sciences in Paris, did not agree, and a feud developed which lasted for years. Brown was also one of the leaders in the movement to reform the Royal Society, by electing only eminent scientists to the fellowship.

After 1828 Brown published comparatively little himself, although he frequently assisted others who were preparing their research for publication. In the museum he was mainly occupied with expanding the collections. The funds available to him for this purpose were hopelessly inadequate. He regularly had to apply for special grants, through the trustees, sometimes successfully, sometimes not, so that important material went elsewhere. His advice was also influential, often decisive, when botanical appointments were being made. He seems to have refused the offer of a civil list pension, although he had no entitlement to a retirement pension from the trustees of the museum. Brown suffered from ill health most of his life. He certainly had an obsessional personality, perhaps some signs of manic depression. He died in his home at 17 Dean Street on June 10, 1858, and was buried at the cemetery of Kensal Green, leaving the large estate of £12 000. The collections Brown had worked on were dispersed.

His personality is difficult to assess, since there are so many conflicting anecdotes regarding the extent of his reserve, sociability, humour and so forth. However he appears to have been a man of unvarying simplicity and benevolence of character who tended to be shy and reserved in the presence of strangers, but warm-hearted and occasionally amusing when in the company of friends. Other distinguishing features were his honesty and judgement, which made him an invaluable counsellor to those who sought his advice. One of his German disciples wrote of 'the prospect to sit down before you, to hear your lovely voice. The voice of my greatest teacher, my adored master, my friend.' His doctor wrote: ' I never presumed to be able to estimate Brown's eminent merits as a man of science, but I knew vaguely their worth, I loved him for his truth, his singular modesty, but above all for his more than woman's tenderness. Of all the persons I have known, I have never known his equal in kindliness of nature. He was the most faithful, the tenderest of human beings . . . '

JOHN JAMES AUDUBON (1785–1851)

To millions of Americans Audubon is the patron saint of birds, a fearless frontiersman who penetrated hitherto unexplored territories, who lived with Indians and drew the birds and animals he loved so well. His monumental work *The Birds of America* is one thing, the slaughter of wildlife, often hundreds of specimens to make one drawing, is another. As a human being he had as many defects as good qualities, as we shall see. He was neither honest nor fair in acknowledging the contributions that others made to the success which eventually came his way. He was only marginally literate and was indifferent to the truth.

The naturalist's father Jean Audubon was a native of the Vendée in south-west France. Born there in 1744 he married at the age of twenty-six an amiable and cultivated widow named Anne Moynet, nine years older than he was; she too was of French nationality. He came from a seafaring family and began making voyages to the French part of the island of Santo Domingo (now Haiti) to bring sugar, coffee and cotton back to France. When the war was over he settled there, making a small fortune as merchant, planter and slave-dealer, while his wife remained patiently behind in Nantes. He died in 1818 having lived through not only the turbulence of the French Revolution and its aftermath but also the American War of Independence, in which France was involved and contributed much to the eventual American victory. A child of his time he hated the British, who seized and imprisoned him for several years.

The future naturalist claimed to be the Dauphin, eldest son of King Louis XVI of France and Marie Antoinette. There is no doubt that the Dauphin died on June 7, 1795 while in prison. In fact Audubon was not the Dauphin but the illegitimate son of his father and a creole named Jeanne Rabin. He was born on April 26, 1785 at Les Cayes, Santo Domingo, and given the full name Jean Jacques Fougère Audubon. His father returned to France with two black slaves to live a few miles downstream from Nantes and operate a coastal vessel, leaving his son to be brought up by his doting

wife, who arranged that he was given some private tuition in mathematics, geography, music and fencing. Encouraged by his father he developed an interest in natural history and drawing, but was never able to write French or English correctly.

In 1803 his father, who had invested some of his savings in a house near Philadelphia, decided to send him to live there, to avoid him being conscripted into the French army. The house, called Mill Grove, was occupied by a tenant. In the months he spent at Mill Grove Audubon began to display certain qualities that never left him: a hunting skill, an undisciplined curiosity, a latent artistic power and an inexhaustible energy. Living nearby was an immigrant family named Bakewell, acquaintances of Erasmus Darwin, Joseph Priestley and other English scientists. Their daughter Lucy became Audubon's long-suffering wife in 1808, after an engagement lasting four years, during which her fiancé was back in France. One of her brothers described him as an admirable marksman, an expert swimmer, a clever rider, a good fencer and possessed of prodigious strength. He was 'notable for the elegance of his figure and the beauty of his features', dressed well, was musical, danced well, and knew various handicrafts.

In the summer of 1807 Audubon and a male companion set off into the interior of the country to make their fortunes. When they reached Louisville, Kentucky, on the Ohio River, they opened a store. Audubon used his wife's dowry to buy stock but took little interest in the business, which was not a success. He returned to Mill Grove to claim Lucy and bring her back to Louisville. For some unknown reason he and his partner moved the store downstream to a smaller place while Lucy and son returned to her parents' home. Finding that business was worse than in Louisville they tried what turned out to be the least successful of his enterprises, a sawmill. This lost Audubon the rest of his own money as well as the money of several other local investors, who naturally blamed him for its failure. He fled to Louisville where his creditors put him in gaol until he pleaded bankruptcy.

Audubon is generally regarded as an artist–naturalist. He was an able draughtsman and sometime in the next few years he learnt to paint in oils. (There is an unlikely story that he studied under the painter Jacques-Louis David.) On his release from prison he made a living by drawing portraits of the inhabitants of Louisville. He also started an art school, but this was not a success. Instead he moved to Cincinnati where he found employment as taxidermist at a museum. Not in the least discouraged by this succession of failures he decided he would produce for publication a comprehensive and complete collection of pictures of all the American birds, life-size and in their natural surroundings. In furtherance of this scheme he travelled through the southern states, down the Ohio and Mississippi Rivers to New Orleans, after which he planned to go east to the Florida Keys and then back to Cincinnati via Arkansas and Hot Springs. In New Orleans he collected a few commissions for portraits but after five months decided to return to Lucy. On the way he landed a job teaching drawing to the attractive but sickly daughter Eliza of a plantation owner. The surrounding countryside was remarkably lush and full of birdlife. It is there that he developed his distinctive style of ornithological illustration. When his appointment was terminated prematurely – he became infatuated with Eliza – he went back to New Orleans where Lucy and their two sons reluctantly came to join him; it was over a year since he had last seen them. Soon after they arrived he left them and went up river to Natchez, where he found a teaching job.

Meanwhile Lucy and the boys also moved up to Natchez where she obtained a post of governess. After a time she was appointed head of a private school for girls which she ran successfully for the next seven years. Meanwhile her husband was wandering from place to place, all the time making drawings of birds for the book he was planning to get published. After several years of this itinerant life he lost hope of trying to interest an American publisher in his work and decided to see if he could do better in England. In 1826 he set sail for Liverpool, armed with letters of introduction which soon bore fruit. The tall stories he told of his life as intrepid explorer, artist, huntsman and naturalist were found fascinating.

His drawings of American birds were exhibited successfully and he began to feel optimistic about getting his work published. Although London was the best place for publishers he first went to Edinburgh, where he was received enthusiastically

Woodcuts usually provided the illustrations for books of all types until copper etchings became more usual in the eighteenth century. The lavishly illustrated books for which there was a market among wealthy collectors were usually hand-coloured until the turn of the century when lithography was invented. In Edinburgh Audubon found an engraver–printer willing to produce the necessary aquatints, a relatively new process. The first batch of plates was produced, and with these to show publishers he set off for London. There he heard that his colourists in Edinburgh had gone on strike and production of *The Birds of America* had come to a standstill. However, at last he managed to find a publisher. There was a shop in Newman Street, just north of Oxford Street, where the Havell family sold art materials and pictures of all kinds and had for several generations practised the art of engraving. This firm undertook to publish *The Birds of America* for him in instalments. When the last of them eventually appeared, in 1838, there were altogether 435 plates and 1065 figures.

Audubon now needed to find subscribers, but unfortunately he had arrived just at the end of the London season when many potential subscribers had already left London for their country estates. So he went to Paris where he met Cuvier and the great botanical artist Redouté who introduced him to the duke of Orleans, later King Louis-Philippe. The duke and duchess both became subscribers. When he returned to London he had added fourteen more subscribers to his list. Before leaving England he entrusted the general supervision of his business to the physicist and naturalist John George Children. Children, who was secretary of the Royal Society, arranged for the king to be shown the first instalment of *The Birds of America*; he not only became a subscriber but also allowed his name to be entered as patron.

Audubon tried to persuade Lucy to come over and join him but after her previous experiences she refused until she was sure he could provide her with financial security, otherwise she insisted he should come to her in Louisiana if they were to be reunited. So in 1829, when he had been away three years, he returned to America. He missed the American countryside and did not greatly care for London. He also wanted to gather as much new material as possible for future numbers of his magnum opus. After joining Lucy in Louisiana he took her with him to Washington, where he obtained another patron in the distinguished Bostonian Edward Everett, then leader of the House of Representatives. This led to an invitation to the White House by President Andrew Jackson. He started writing his three-volume *Ornithological Biography*, an account of his life in which imagination and exaggeration play an important part.

Copies of the first edition of *The Birds of America* were sold for $1000. Both of Audubon's sons were involved in the business of production and distribution. Victor, the elder son, supervised in London, and in 1834 Audubon returned to visit him. After two years he was in the United States again and visited the Pacific coast for the first time. Then back to London for another three years to supervise the publication of the *Ornithological Biography* and the production of a small-scale and much cheaper version of *The Birds of America*, which became a best-seller. Three years later he returned to America a wealthy man. He bought thirty-five acres of land on the Hudson River in what is now the Washington Heights area of New York City, where he built a spacious wooden house with a high portico running the length of the side facing the river. He worked on his last book, *The Viviparous Quadrupeds of North America*, which was in the same grand style as his magnum opus, but much less important. He made another journey to the West, exploring the little-known region of the upper Missouri and Yellowstone Rivers. Soon after this his mind began to fail, and he died on January 27, 1851.

SIR WILLIAM HOOKER (1785–1865)

Both the botanical Hookers, often known as the Hookers of Kew, will be profiled here (the younger in a later chapter). William Jackson Hooker was born in the city of Norwich on July 6, 1785. His father, a merchant in the wholesale woollen trade, was not a wealthy man but he cultivated rare plants as a hobby and liked reading books about travel. In addition to what was taught at Norwich grammar school William's education included some experience of farm management, since when he came of age he would inherit some landed property from a deceased godfather. While still quite young he decided to devote his life to travel and natural history. As a result of discovering a botanical rarity near Norwich he came to the notice of Sir James Smith, the purchaser of the effects of Linnaeus. Smith introduced him to Dawson Turner, a wealthy Yarmouth banker and amateur botanist, who was preparing a book on fuci (seaweeds) and asked Hooker to prepare illustrations for it. Impressed by the knowledge and skill of his young visitor Turner gave him an introduction to Sir Joseph Banks, and helped to secure his election to the Linnean Society. Hooker went to London, called on Banks and met his resident librarian Drysander. Meanwhile Turner took

the young man along on a family visit to the Scottish highlands in 1808. The next year Hooker returned to Scotland on his own and formed a plan to venture much further afield. With the encouragement of Banks he arranged to join a vessel that was bound for Iceland. On the return journey the ship caught fire. Although those on board were rescued all the material Hooker had collected was lost.

By this time Hooker should have come into his inheritance but there was an ongoing lawsuit which delayed matters. Like Banks himself he expected to be able to pay his way on a major voyage to the tropics. Banks proposed Java as a suitable destination, but Hooker's parents raised objections, because of the risk of catching malaria, and Hooker decided not to go. Instead he accompanied the Turner family on a year-long botanical tour of France and Switzerland, meeting the great naturalists von Humboldt, de Candolle and Bonpland, who were impressed by his botanical knowledge. A few months later he married Maria, the eldest of Turner's daughters; she bore him one son and two daughters. The banker himself was in some financial difficulty because some of his customers had defaulted, and this rebounded on Hooker, who found he was no longer wealthy. To provide financial security for his growing family, with two more children to come, he applied successfully for the vacant Regius professorship of botany at Glasgow University.

Although Hooker was a great success at Glasgow his ambition was to become director of the Royal Gardens at Kew. Because this had fallen into neglect after the death of Banks in 1820 a Treasury Commission under John Lindley (1799–1865) was set up to consider its future. The key recommendation of the Commission was that it should be developed into a national botanical garden. The administration of the gardens was reformed and Hooker was appointed director. Lindley, a protégé of Hooker's, was also a strong candidate; he was the first professor of botany at London University and much else. The gardens then consisted of just eleven acres, but in five years Hooker had extended it in stages to the 650-acre Royal Botanical Gardens we know today and established its main departments. He added several plant houses and he also opened the gardens to the public, on a limited basis, and created an efficient system of exchanges of plant material with botanical gardens elsewhere. There was considerable rivalry with the natural history department of the British Museum. Where Kew scored was by acting as a nursery for plants of economic importance and supplying them to parts of the British Empire where they might flourish. This role was promoted in the Museum of Economic Botany, created by Henslow and Hooker, where the public might see plants 'eminently curious or in any way serviceable to mankind'.

Although when young Hooker had hoped to visit the tropics he never did. His many articles and books were written and often illustrated with the help of the vast private herbarium he had built up at Glasgow. In his lifetime this was made available to botanists and after his death was purchased for Kew by the nation. There was disagreement at this time as to whether the national herbarium should be at the British Museum, as Brown advocated, or at the Royal Botanic Gardens, as Hooker preferred; eventually this was settled in favour of Kew. William Hooker died on August 12, 1865, at the age of eighty. His wife, who for sixty years had acted as his secretary and amanuensis, died in 1872. He was elected to the Royal Society in 1816, and knighted in 1836, for his services to botany. He maintained an enormous correspondence with botanists and others throughout the world, and was much honoured later in life. As we shall see in Chapter Six, he was succeeded as director of Kew Gardens by his son, who had been appointed assistant director ten years previously.

5   From Gould to Darwin

JOHN GOULD (1804–1881)

The next profile is of a man who used his fascination with birds as a way to make money. His father, also named John Gould, was one of ten children who came from a farming family in the county of Somerset. John Gould senior had moved away from his birthplace to take up a job in horticulture at the seaside town of Lyme in Dorset, where he married Elisabeth Clatworthy in 1803, and the son she bore him a year later is the subject of this profile. By the time he was fourteen his father had become foreman in the royal gardens at Windsor, and it was there that Gould somehow learnt the art of taxidermy. Since it was assumed he would want to follow in his father's footsteps he was sent to learn something about it at Ripley Castle in Yorkshire before returning to Windsor where the king, George IV, encouraged him to develop his taxidermic skills, and of course royal patronage led to others seeking his services.

In 1827 he participated in a competition held by the newly formed Zoological Society in London, in which he was easily the first, and as a result at the age of twenty-three he was appointed the society's first curator and preserver of specimens. At this time the vast zoological collections of the British Museum were in a sad state of neglect, and new material was arriving all the time, so that there was plenty of work for Gould. He also became known to the general public when the king called on him to help stuff a young giraffe which died not long after it was brought from its natural habitat to live in the royal menagerie in Windsor Great Park; it had become a special favourite of the king.

Gould had always been fascinated by birds and now saw a way to profit by this. Audubon had shown there was a market for lavishly illustrated books about birds. New species were being discovered all the time as the less accessible parts of the British Empire were explored. Gould's first published work, his *Century of Birds from the Himalaya*, based on a collection of bird skins the Zoological Society had received towards the end of 1830, contained 80 plates figuring 102 birds.

Gould had a good head for business but was no draughtsman. He solved the problem by marrying a competent artist named Elizabeth Coxen in January 1829 and shortly afterwards securing the services of a dedicated and dextrous secretary, Edward Prince, who served him loyally for forty-five years. The part these played in his later success should not be underestimated. Determined, intelligent, educated, practical and obedient Elizabeth was the first of several artists who turned his crude sketches into the beautiful pictures to which his name is attached. Although the true artist was acknowledged, and it was never Gould himself, he was happy to acquire an undeserved reputation as a bird illustrator. Of the many people who worked for the Gould firm over the years the best known is Edward Lear, famous for his watercolour landscapes and nonsense rhymes. Lear began his career in poverty and insecurity as draughtsman. It was he who provided the impetus for Gould's first publication and it was Lear who later transformed Gould's static and rather unimaginative style into the confident and innovative work that characterised his subsequent publications.

Lear had been a lonely awkward boy, frightened by depression, the deaths of four of his sisters, and frequent attacks of epilepsy. He found relief from his fears and inadequacies in a world of fantasy and art. In particular he made a study of natural history, especially of birds, with which he felt a strange affinity. Rejected by his mother at an early age (she had twenty-one children altogether) he was brought up by two of his elder sisters who taught him to draw. Inspired by the work of Audubon, Lear decided, at the age of nineteen, to produce a folio volume of forty-two lithographs of the family of Psittacidae or parrots. This brought him to the attention of the earl of Derby, a keen naturalist, who had formed a unique private menagerie at his country house, Knowsley Hall, near Liverpool, which after his death became the nucleus of the London Zoo. The earl commissioned Lear to produce an illustrated guide to it, and as a result Lear spent the next four years painting at Knowsley and living as a friend of the family. Other commissions followed.

Lear was in advance of Gould in many respects but the latter's competition proved too much for him and he sold the copyright of his own book to his rival when it was still incomplete and remained so. Under Gould's auspices they started work on *Birds of Europe*, for which Lear produced sixty-two of the total 448 plates. He chose the most striking subjects, leaving the others to Elizabeth. He continued to contribute to Gould's publications and it was his work more than anything which made them a success. Gould played down Lear's contributions, however, and the latter began to nurse a well-justified sense of grievance. Eventually Lear resigned himself to Gould's impenetrable isolation but long after they parted company he described his former employer and erstwhile friend as 'a more singularly offensive mannered man than hardly can be but the queer fellow means well, tho's more of an egoist than can be described'.

Two of Gould's brothers-in-law who had emigrated to Australia were sending him bird skins from there. Gould realised that the largely unexplored continent held splendid opportunities for the discovery of new and exotic species. He decided to go there himself, leaving Prince in charge of the firm. To whet the appetite of connoisseurs for the more ambitious *Birds of Australia* he had in mind he brought out *A Synopsis of the Birds of Australia and the Adjacent Islands.* He took his wife along too, but left behind their children in the care of relatives, much to her regret. His party left London in May 1838 and arrived in Hobart, the chief town of Tasmania, after an uneventful journey of four months. On the way he was able to study many types of seabird. Unlike many of the settlers he was attracted by the aborigines, who served him well as guides and who shared his interest in birds and other creatures.

After five months on the island he had collected 500 specimens of the birds of Tasmania, including at least fifty he thought new to science, and left for continental Australia itself. He obtained advice from the explorer Charles Sturt, an expert ornithologist who had made a large collection of watercolours of birds he had found on his various expeditions. Gould at once offered a high price for them but Sturt refused to sell. Shortly afterwards they were stolen. Meanwhile his homesick wife, left behind in Tasmania, had given birth to another child. Gould was off to Adelaide where he found a multitude of fine specimens. Next he returned to Hobart to collect his wife and children and arrange for his Tasmanian collection to be shipped to Sydney. From there the couple set off up the coast to Newcastle, his base for a journey some 400 miles into the outback. After four very successful months of collecting they returned to Sydney to prepare for their voyage home. He had taken out from England an assistant named John Gilbert, a younger man with a background similar to his own, who was devoted to his employer's interests. Gilbert had been sent off to collect for Gould independently and the plan was that they should meet in Sydney. However Gould left without seeing Gilbert, who was very short of money and equipment.

Once back in London Gould soon began to organise the production of *Birds of Australia*, a project that was to take eight years in all. It remains the most comprehensive work ever produced on the subject; the vast lavish hand-coloured illustrations, with full descriptions, remain a historical record and basis for research. He received a hero's welcome from the scientific fraternity, and was elected to the Royal Society.

Poor Gilbert, his loyal assistant, received no such recognition when he returned to London, and so he went back to Australia where he joined an expedition to Queensland and the north. The expedition was a success but Gilbert was murdered by aborigines in the course of it. Gilbert spent five and a half years of collecting in Australia, as opposed to Gould's eighteen months, visited each of the present states of the continent, and discovered sixty or seventy new species for Gould to name.

Elizabeth died in 1841, at the age of thirty-seven, a few days after giving birth to their eighth child. Gould bore his loss with fortitude. He immersed himself in his work, to which she had contributed so much, and her place was difficult to fill. He recruited a minor artist named Henry Constantine Richter, who grew accustomed to the task of transforming Gould's crude sketches into proper illustrations, but they are the works of a draughtsman rather than an artist.

As well as *Birds of Australia* there were other projects, such as Gould's monograph on the kangaroo. He offered a major portion of his collection of specimens to the British Museum, which had already bought some of them, and was indignant when the trustees turned him down. Eventually they went to a museum in Philadelphia. Gould had a lifelong obsession with hummingbirds. By the time of his death he had amassed a collection of 1500 mounted and 3800 unmounted specimens. His greatest achievement was the publication of a five-volume monograph of the Trochilidae, or family of hummingbirds, begun in 1849; his vast array of specimens proved one of the great attractions of the Great Exhibition of 1851. After this closed his exhibit was transferred to the Zoological Gardens.

In 1857 Gould fulfilled a longstanding ambition to visit America. He had never yet seen a living hummingbird but these were plentiful around Washington. He brought some back to England but the British climate proved too much for them. However the trip was considered a success in other respects. In 1859 Gould moved from Broad Street, where he had lived and worked for nearly thirty years, to the more salubrious neighbourhood of Charlotte Street. He began work on the long-awaited *Birds of Great Britain*, for which he recruited the services of a young German zoological painter named Joseph Wolf, later to be described as the greatest of all animal painters. He was so good at depicting birds that Edwin Landseer remarked 'he must have been a bird before he became a man'. Gould appreciated Wolf's talents, whereas he had so deprecated Lear's. In 1854 a cholera epidemic swept the capital; one of the victims was Gould's eldest son. In 1873 he lost his youngest and most favourite son who died of fever on board ship in the Red Sea. After this Gould began to suffer from ill-health, as did his invaluable assistant Price. Gould withdrew into himself, retreating into a private world of neurosis and self-obsession. He died on February 3,1881 at the age of seventy-six.

Gould had distinguished himself during his long, hard and ambitious career as a pioneer in the new science of ornithology. He was one of the foremost collectors in the world, a first-rate taxidermist and taxonomist and one of the finest publishers in London. A mean, sometimes unscrupulous businessman, he made a fortune out of his work. Gould was slighted by the Cambridge-based ornithologists and was never invited to join the British Ornithologists Union, formed in the 1850s. A friend of his remarked that he had a really tender and affectionate heart hidden though it was beneath a highly sensitive reserve, which never permitted him the relief of expression. 'Gould's industry, enthusiasm and perseverance were beyond praise', said Wolf, 'but he was a shrewd old fellow and the most uncouth man I ever knew.' When Lear heard of Gould's death he reminisced: 'He was one I never liked really, for in spite of a certain jollity or bonhomie he was a harsh and violent man. Ever the same persevering hardworking toiler in his own line, but ever as unfeeling for those about him. In the earliest phase of his bird-drawing he owed everything to his excellent wife and to myself, without whose help in drawing he had done nothing.'

SIR RICHARD OWEN (1804–1892)

Richard Owen was born in Lancaster on July 20, 1804, the second of six children and the younger of two sons. His father, Richard Owen senior, was a West India merchant; his mother, born Catherine Parrin, was of Huguenot descent. The father died when his son was five, just before he started formal education at the local grammar school. After ten years there, where he was reported to be lazy and impudent, he became apprenticed to local surgeons–apothecaries, thus gaining access to post-mortem dissections at the county gaol. Four years later he matriculated at Edinburgh University but stayed only half a year before moving to St Bartholomew's Hospital in London, where he was appointed prosector to the lectures of John Abernethy. The following year he became a member of the Royal College of Surgeons and set up in medical practice, at the age of twenty-one. A year later Abernethy arranged for him to become assistant conservator at the Hunterian Museum to assist Clift in the huge task of cataloguing the collection.

Owen was six foot tall, rather gaunt-looking with an immense head, with high cheekbones, a wide thin-lipped mouth, a prodigious forehead and very large eyes. His hands were large and supple. Music was his main recreation: he could sing and play the cello and he regularly visited both theatre and concert hall. When Weber's opera *Oberon* was first performed at Covent Garden, apparently Owen attended all thirty-one performances. He was a first-class chess player, and he had many acquaintances in the artistic and literary world of his day. Yet 'this eminent and highly gifted man can never act with candour or liberality,' it was said. The negative side of his character was experienced mainly by his peers; to students and people he only met socially he only showed the positive side of his nature. Owen was a deeply religious person who belonged to the Established Church.

Owen's employment by the Royal College lasted almost thirty years. However, after the first year he added a lectureship in comparative anatomy at St Bartholomew's Hospital. Also he began dissections of animals that died at the Zoological Gardens. He now started publishing, first a memoir on the pearly nautilus, then contributions to an encyclopaedia, and other articles.

In 1834 he was promoted to the newly established chair of comparative anatomy at St Bartholomew's. He was also elected to the Royal Society, at the age of thirty, for his work on marsupials and monotremes.

Owen was now ready for marriage; his bride was Caroline Clift, daughter of the late conservator under whom he had worked at the Hunterian. They had been engaged for nearly eight years, probably because of his inadequate and insecure income. Their home for the next seventeen years was an apartment in the building of the Royal College of Surgeons. About this time he let his medical practice lapse, having decided that he would henceforth devote himself to the advancement of science. He was also elected Hunterian professor two years later; he was immensely proud of this. By this time his work had achieved international recognition; he was elected a corresponding member of the Académie des Sciences in Paris and awarded other honours. He succeeded Clift as conservator of the Hunterian Museum. Increasingly he was called upon to serve on commissions and committees, especially those to do with the future of the Hunterian and the natural history collections of the British Museum, of which he became a trustee as well as superintendent of the collections. He was elected to the prestigious Athenaeum under a special procedure which avoided the necessity of balloting the membership-at-large, and he also belonged to other clubs, which brought him into social contact with men of influence and power; later Owen arranged the election of the ornithologist John Gould in the same way. Owen was increasingly in dispute with the Museum Committee of the Hunterian Trustees about some matters, with the Council of the Royal College about others. When he was offered the prestigious three-year Fullerian professorship at the Royal Institution, against strong competition, he was obliged to decline it until he ceased to be Hunterian professor, when the offer was renewed.

In 1830 Cuvier visited the Hunterian Museum; a year later Owen made a return visit to Paris, where he was impressed by Cuvier's Museum of Comparative Anatomy and from then on his ambition was to merge the Hunterian collections with those in the British Museum and create a national collection of comparative anatomy comparable with the one in Paris. Meanwhile it was agreed to expand the facilities of the Hunterian Museum by demolishing the old building and constructing a new one on the same site with more exhibition space. However the ruling clique of the College wished it to remain a purely medical museum, catering for the education of surgeons, while the aristocratic trustees of the British Museum wished to keep all their various collections close to the national library, as Sloane intended. The Darwinians, led by Huxley, agreed with the latter. Huxley, of course, not only defended Darwin but lost no opportunity to attack and discredit Owen, as we shall see when we come to his profile in a later chapter. Owen began to think that it would be better if the Hunterian collections were transferred to the British Museum, rather than vice versa, but the Royal College would have none of this. In 1851 he was granted by Queen Victoria the use of a grace-and-favour residence, Sheen Lodge in Richmond Park, where he lived for the remaining forty years of his life. This honour caused further trouble with the council of the Royal College, who thought he should live in central London, and offered him a house in Lincoln's Inn Fields instead. In 1856 Owen resigned from his Hunterian positions on being appointed superintendent of the overcrowded natural history departments in Bloomsbury. Almost at once he began a campaign, eventually successful, to detach natural history from the British Museum. Twenty-five years later his ambition was realised when the new Natural History Museum, designed by Albert Waterhouse, was opened in South Kensington, with Owen as director, subject to a Board of Visitors but without the need to report to a Board of Trustees. Like the younger Hooker at Kew, Owen developed a network of contacts who supplied him with additions to the museum's collection of specimens. Expeditions were given specific instructions of what to look for, although Owen never went on one himself.

He was an excellent lecturer, not only to the captive audiences he taught as Hunterian professor, but at the Royal Institution during his stint as Fullerian professor and at the Royal School of Mines, which lent him its lecture theatre when he was at the British Museum, where he attracted a predominantly middle-class and professional audience. He also lectured at the literary and scientific societies of various places in the provinces. He fascinated his audience by describing those extinct monsters for which he coined the name of 'dinosaurs'. Much of the prestige attached to comparative anatomy lay in its successful application to the identification of extinct vertebrates, such as the *Megatherium* or giant ground sloth, from partial skeletons. Unlike Darwin he was not deeply interested in geology, but because he was an expert on fossils he was three times invited to become president of the Geological Society.

Owen's views on the origin of species underwent substantial change in the course of his life. Although he did not necessarily accept traditional interpretations of what was written in the Bible he was reluctant to accept any scientific theory that seemed to challenge religious belief about the origin of man. During the 1830s he was attracted by the theory of multiple acts of creation. By the mid 1840s, however, Owen had become convinced of the fact of organic evolution, but uncertain of the cause, although sure that it was by natural laws rather than through divine intervention. After the meeting of the Linnean Society at which Darwin's and Wallace's papers were read, he referred favourably to the idea of natural selection, but a few months later he was urging caution. When the *Origin of Species* appeared the following year he was appalled to find that Darwin cast him, like Agassiz, as a leading advocate of the immutability of species. In later editions Darwin corrected this but still misrepresented Owen's position.

Owen's wife Caroline died in 1873 and their only child William mysteriously committed suicide in 1886 at the age of forty-eight, jumping into the Thames while leaving his hat with purse, watch and address card inside it on the riverbank. A biography of Owen was written by his grandson Richard Starton Owen. Owen himself suffered from deafness and stomatitis; after a lengthy decline in health he died of old age at Sheen Lodge in 1892. During an active career spanning nearly six decades almost every honour that could be bestowed upon a scientific figure came his way; the Wollaston, Royal and Copley Medals, a civil list pension, a knighthood on his retirement, honorary doctorates from Oxford, Cambridge, Dublin, Edinburgh and very many more.

By all accounts Owen was a sociable person with an active interest in the fine arts of his day. To outsiders he was always most charming and instructive in conversation, helpful to colleagues who posed no threat. However in relation to his peers his power of hatred was extraordinary. He was described as not only ambitious, envious and arrogant but also untruthful and dishonest. Darwin, in particular, described him as spiteful, unfair, ungenerous and extremely malignant. Others accused him of being false, rude, unjust, illiberal, disingenuous, capable of gross misrepresentation, abuse, bitter sneers and mean conduct. The clash with Darwin has tended to overshadow Owen's very real achievements, which led to him being compared with France's Cuvier and Germany's Alexander von Humboldt. He wrote over 600 detailed monographs and scientific papers, which made known many organisms both recent and fossilised, helped to delineate several natural groups, and laid the bases for much later work by others. He should be particularly remembered for his achievement in bringing about the foundation of the separate Natural History Museum in South Kensington, where a bronze statue of Owen was placed in the main hall.

LOUIS AGASSIZ (1807–1873)

Jean Louis Rodolphe Agassiz, the son of the Protestant pastor of Motier-en-Vuly, in west central Switzerland, was born on May 28, 1807. His strong-willed father Rodolphe came of an old Vaudois family, noteworthy figures in the cultural life of the prosperous region. His mother, Rose Mayor, was the daughter of a physician who practised in a nearby village; her family was also eminently respectable and influential in that part of Switzerland. She was a woman of strong convictions, remarkable intelligence, a keen wit and an energetic disposition. After the birth of Louis, the first of her children to survive infancy, she presented her husband with three more children, a son Auguste and two daughters Olympe and Cécile.

At the age of ten Agassiz was sent twenty miles away from home to begin his formal education at the college of Bienne, where he shone at languages, but not at mathematics and the physical sciences. At the end of five years he began to plan the next stage in his life. In the town of Neuchâtel he had an uncle, François Mayor, who directed a prosperous commercial house, and it was hoped in the family that he too might succeed in commerce. Louis was soon to demonstrate that he did not have a head for business. In fact he had already become obsessed by natural history and resolved to dedicate himself to the world of science. With characteristic determination he set

about transforming himself from a conventional rural youth into a mature scholar. He planned a period at a German university, followed by further studies in Paris. To please his father, however, he agreed to begin with two years at the Academy of Lausanne, a college which prepared students for entrance to the University of Zurich. It was also agreed that he would study medicine at Zurich, but still it was natural history that absorbed most of his time and energy.

In the German tradition it was quite usual to attend two or three medical schools before qualifying. He chose to move from Zurich to Heidelberg for eighteen months and then from Heidelberg to Munich. He assured his father that after that he would obtain his medical degree, return to his native Switzerland and begin to practise as a physician. After three years at Munich he passed his medical examinations without effort, while completing his Ph.D. thesis, a moderately distinguished, precise description of a collection of fishes indigenous to Brazil. This was published as a book, which he dedicated to Cuvier, the dean of French natural science. The next step was to Paris, at the end of 1831, where he presented his work to Cuvier, who was greatly impressed by what the charming young Swiss had achieved. Both were children of the Jura, but Cuvier was rather solemn, while Agassiz was always smiling, often laughing. Both possessed an extraordinary memory. Agassiz became a disciple of Cuvier, adopted his principles of classification, and followed him in refusing to accept the evolutionary theories of Lamarck and his school. For the rest of his life Agassiz believed, like Cuvier, that an all-powerful deity had planned the entire range of past and present creation, making impossible any genetic connection between ancient and modern forms.

Alexander von Humboldt, who was in Paris in an official capacity to report on the government of Louis-Philippe to his Prussian sovereign, took Agassiz along to the Collège de France to hear the debates between Cuvier and Geoffroy. When Cuvier died the following year, as a result of the cholera epidemic, Agassiz, feeling the need for a patron, turned to Humboldt. The wealthy and influential Humboldt enjoyed discovering talent and offering a word here, a letter there and occasional financial help. Thanks to him, Agassiz could finish the work he had planned in Paris, but after that he needed a secure position that would provide him with a steady income. There was a chance that something might become available in the French capital, but then he heard that there were plans to establish a museum of natural history and a new academy in Neuchâtel. The canton and town of Neuchâtel had been under Prussian rule for over a hundred years, apart from one short break; the King of Prussia was also Prince of Neuchâtel. He had a policy of supporting cultural activities of which he approved. Humboldt was pleased to use his influence with the king, through his brother Wilhelm, to ensure that Agassiz was appointed professor of natural history in the academy, which was inaugurated in 1835.

In 1832 Agassiz returned to his homeland as an independent scholar, supported by the Neuchâtel aristocracy and the Prussian monarchy. The next fourteen years were the most productive of his career. Hoping to find examples of separate creation he spent several years studying the fossil fish collections in museums and private holdings throughout Europe. This project originated before he first went to Paris to see Cuvier. There he had learned that Cuvier himself had been preparing a book on the same subject, left unfinished because of Cuvier's death. Agassiz took over the material Cuvier had assembled, and added his own research. The result was a masterly series of six monographs entitled *Recherches sur les poissons fossiles*, in which more than 1700 ancient fishes were analysed by the comparative method first taught by Cuvier. This won high praise from authorities such as Lyell and Owen, the English pioneers of geology and palaeontology, and was the basis of Agassiz's international fame and scientific fortune.

Agassiz had no business sense: he was too impatient and susceptible to flattery. So that his monographs and other works should be published as lavishly as he thought they deserved, he established a kind of publishing factory in Neuchâtel by taking over a local printing firm. He gave inspirational talks, rather than clear directions, to his assistants who were given board and lodging at his house and paid their expenses instead of a reasonable salary. Sometimes he gave them work for no other reason than to keep them busy. He started producing translations of British books on natural history, without asking their authors' permission or even making his intentions known to them. The high-handed young man could not see anything unfair about such piracy, maintaining that the English editions of such useful works were inaccessible to continental scientists on account of their great cost. During the fourteen years of its existence the business produced more than twenty volumes, with two thousand plates, and many separate papers; all were well written, beautifully printed and profusely illustrated. However they sold at a loss and he found himself heavily in debt. Eventually he was forced to close the factory.

First at Heidelberg and then at Munich Agassiz had enjoyed the close friendship of a classmate, Alexander Braun, whose home was in Carlsruhe. Alexander had a sister Cécile (not to be confused with Agassiz's sister Cécile) to whom Agassiz proposed marriage. She was a capable artist who would be useful to him professionally. They were married in 1833; she bore him three children: a son Alexander, born in 1835, and two daughters, Ida, born 1837, and Pauline, born 1841.

At the age of twenty-one he had predicted he would become the first naturalist of his time, and it was beginning to look as if he might be right. 'I feel within myself the strength of a whole generation to work toward this end,' he maintained. Someone who knew him well during this period wrote this description of him: 'Agassiz was one of the most brilliant men of his time. Young, handsome, of an athletic constitution, gifted with a captivating eloquence, his spirit was animated by an insatiable curiosity, his memory excellent, his perspicacity rare and very keen, and his way of judging and coordinating facts highly philosophical in its tendency.'

Agassiz was squarely built, with broad shoulders and a powerful and well-proportioned body, of rather more than average height, with unusually large but well-formed hands but comparatively small feet. His head was simply magnificent, his forehead large and well developed; and his brilliant, intelligent and searching eyes could best be described by the word fascinating, while his mouth and somewhat voluptuous lips were expressive, and in perfect harmony with an aquiline nose and well-shaped chin.

In 1834 Agassiz made his first visit to England, where he was received by his fellow naturalists with open arms. He was impressed by the splendid collections of fossils and living animals he was shown, and by developments in the Zoological Gardens. In 1836 he was awarded the Wollaston Medal of the Geological Society of London, a unique distinction for a young naturalist, after he had demonstrated that the large boulders found on the sides of alpine valleys had been carried there by glaciers. In fact Agassiz was not the first to advance this theory, but he understood its significance better and developed it further. In 1837, in an address given at Neuchâtel, he discussed the effects of glacial action on topography, and maintained that in geologically recent times widespread glaciation had formed an Ice Age. The evidence for this was published in his *Etudes sur les glaciers* of 1840. After visiting the Bernese Oberland for the first time, he decided to set up a research station on the edge of the Aar glacier. This cabin, under a huge rock, became known as the Hotel des Neuchâtelois (sometimes misguided tourists tried to book rooms in what they took to be a proper hotel). On a visit to Scotland with the British geologist William Buckland (1800–1874) he found further evidence of glacial action. Naturally the catastrophists welcomed the idea of recurrent ice ages, which must have destroyed much plant and animal life.

Meanwhile Cécile was becoming increasingly unhappy. She was homesick for Carlsruhe, with its environment of art, music and cosmopolitan culture, quite lacking in Neuchâtel. She also took a strong dislike to Agassiz's secretary Edouard Desor, urging that he should be dismissed, but her husband regarded him as indispensable. She decided to leave her husband and return home to her parents, taking with her their daughters Pauline and Ida, but leaving their son Alexander behind to finish his school education in Neuchâtel.

Agassiz had for some time been planning an American lecture tour, having been promised substantial lecture fees, which he could use to reduce his debts. Influenced by Humboldt, the king of Prussia granted him $3000 for a two-year study comparing the flora and fauna of the United States with those of Europe. After various delays Agassiz left for America in the spring of 1846 on what was intended to be a temporary visit. On the way to Liverpool to catch his ship for Boston he called at Carlsruhe to say goodbye to his wife, who was in poor health, and his two daughters. He also broke the journey in Paris, to get up to date on new scientific discoveries and to defend his theories before an audience of savants. He also renewed his acquaintance with the rich holdings of the Jardin des Plantes, as well as the numerous private collections in the French capital. At this stage he would have been happy to accept the offer of a permanent position in Paris, but it came as rather a relief to some of the savants that this rather overwhelming young man did not settle down among them. Similarly in England, although the leading scientists were friendly enough, and most interested in his discoveries. He had previously declined offers of positions in Lausanne and Geneva. As it turned out the rest of his career was to be in the United States.

The Americans welcomed Agassiz with open arms. Huge audiences came to his lectures, although he was not yet at home with the English language. From his base in Boston, he travelled to the principal cities of New England to meet fellow scientists and to give lectures, and he also made field trips. He was impressed by the vigour and enthusiasm of the people he met and by the excitement of a new land and culture. The tour was a success financially as well, so that he was able to pay off the debts he had incurred at Neuchâtel and bring over some of his assistants. One of these was his secretary Desor who knew nothing about natural history originally, but had learnt so much while working for Agassiz that he now saw himself as an equal rather than a subordinate. Somehow his demeanour had changed and he behaved in various unacceptable ways. For example he spent the fees that Agassiz earned by lecturing on bringing over from Europe not only more assistants but also two Swiss servants and a German gardener. Eventually matters came to a head and Desor was dismissed, after a good deal of trouble.

Although Agassiz had only intended to leave Europe temporarily the great opportunities for studying his favourite subjects, and the financial support available in America, induced him to settle there. In 1847 he was appointed professor of zoology and geology with tenure at Harvard and launched into ambitious programmes of research and teaching. Through his popular lectures to laymen he very soon became a public figure. His research output was still Herculean but except for a kind of summer school on the northern shore of Lake Superior he did not travel far into the interior of the country at this stage. Cécile, in Carlsruhe, had been diagnosed with pulmonary tuberculosis and she died in 1848. His son Alexander joined Agassiz the next year; while his two daughters were placed in the care of their grandmother.

Before long Agassiz married again. By far the most distinguished foreigner to visit Boston for many a year, a fascinating conversationalist with excellent manners, he was an honoured guest at the homes of leading Boston families. Among them was the home of the Cary family, where he met an attractive, dark-haired, graceful, intelligent woman of twenty-seven named Elizabeth Cabot Cary. He proposed marriage to her and they were married soon afterwards. His new wife made a comfortable home for his children, attended his lectures, acted as his secretary, and accompanied him on his travels. She took a special interest in her stepson Alexander, who was showing scientific ability and became one of Agassiz's many research students.

Increasingly the Americans came to realise what a valuable asset Agassiz was to American science generally, not just natural history. He was appointed one of the regents of the Smithsonian Institution, in which he took a great interest. He was full of schemes for new research and publications. He made regular visits to Charleston, South Carolina; on one of these he caught malaria and became seriously ill. He made a prolonged tour of the South, travelling up the Mississippi River with its rich fauna. Since scientific books were scarce in the United States at that time he brought his personal library over from Neuchâtel. To provide room for his material Harvard built a large house for him on the corner of Quincy Street and Harvard Street. Alarmed that her household, like Cécile's, might suffer to provide funds for her husband's activities, Elizabeth set up a school for girls of college age in her home. Although the education of young men was well provided for in the Boston area there were no equivalent facilities for young women. The school was a success right from the start, and the fees provided a useful source of income.

In 1854 a letter arrived from Zurich offering him a professorship at the newly established Federal Polytechnic. A short time later a similar proposal arrived from Edinburgh University. These failed to interest him. Three years later Agassiz was offered the professorship of palaeontology at the Muséum d'histoire naturelle in Paris, on very favourable terms. A few years earlier he would have accepted, but he replied that he now felt committed to America, although he recognised that the chair was one of the most brilliant posts in natural science. The French improved what was already a generous offer, and kept it open for two years, but he did not change his mind. Ever since he had been appointed to the Harvard faculty he had been trying to persuade the university to establish a major museum of natural history. Now he could threaten to resign unless he had his way. He was always good at fund-raising and quickly found the cost of the building from wealthy Bostonians and the Commonwealth of Massachusetts. In 1860 the Museum of Comparative Zoology was opened to the public. He planned its exhibits to demonstrate the Creator's master plan for the natural world, emphasising the distinct and separate creation of species. This required that he regarded as distinct species what others regarded as varieties of one species. Soon the museum was overflowing with specimens collected from far and wide.

However, the establishment of the museum was by no means all. In 1855 he had decided to produce a series of illustrated volumes covering the entire natural history of the United States. He persuaded wealthy individuals and institutions to help finance the project by making advance subscriptions. Only four volumes of the series were published; the first was an exposition of Agassiz's creationist philosophy, two others consisted of a technical discussion of the embryology of North American turtles. These were largely written by assistants but Agassiz maintained that work begun under his direction was his intellectual property and needed no acknowledgement.

In 1859 Agassiz, with his second wife and youngest daughter, made what proved to be his last visit to Europe. On the way he renewed his contacts with the naturalists of London and Paris, but the main purpose was the show Elizabeth his homeland and to introduce her to his aged mother. He also purchased some private palaeontological collections and libraries to send back to Harvard. As usual payment was to follow, but the necessary funds had not yet been raised and when they were the onset of the Civil War led to a sharp depreciation in the value of the dollar, causing further difficulties. Agassiz supported the Union cause with an uncompromising patriotism; and it was then that he took out American citizenship. The publication of Darwin's *Origin of Species* in 1859 led to controversy everywhere. In America the leading naturalists became Darwinians, except for Agassiz, who clung dogmatically to the theories of his youth and rejected the idea of evolution of species entirely. He was dismayed when many of his followers turned against his views, but as the dominant publicist in his field his pronouncements had widespread support, particularly among the laity. He seemed less and less to care for the opinions of his scientific colleagues. Professional naturalists commended his encyclopaedic knowledge but not his theorising.

At fifty years of age Agassiz was still full of energy, but his health was not what it was. He led an active social life. He made a lecturing tour in the Midwest, always attracting huge audiences. He made a long-postponed visit to the Rocky Mountains, where he correctly found evidence of glacial action and some interesting fossils, made visible by the construction work in progress on the Union Pacific Railroad. He spent some time at Cornell University, on the way home, and accepted an appointment as non-resident professor.

In 1869 Agassiz delivered a memorable two-hour oration in commemoration of the centenary of Alexander von Humboldt's birth, but immediately after returning home suffered a stroke which left him completely paralysed. In time the paralysis disappeared but a long period of convalescence was prescribed. For the workaholic Agassiz this inactivity was almost unbearable. Eventually he felt ready to resume normal life, although his speech remained affected. A scientific cruise up the Amazon and around the coast of South America was undertaken with his customary enthusiasm. As usual he was accompanied by his wife who afterwards collaborated with him on writing a popular account of their travels. In the Straits of Magellan he found plenty of past and present evidence of glacial action. When they reached the Galápagos Islands they collected yet more specimens, but this was no epiphany for Agassiz as it had been for Darwin. He disembarked in San Francisco feeling too tired to explore the Californian countryside and returned to Boston overland. He had planned to retire when he reached home, handing over responsibility for the museum to his son Alexander, but changed his mind. He went around claiming, on the basis of no real evidence, that the Amazon basin had been covered by glaciers in the past. He now began to plan an extension of the museum to provide room for the huge collection of Brazilian material to be properly housed instead of being piled up in the basement.

When he had recovered from the strenuous journey he turned his attention to his last major undertaking, the establishment of a zoological field station on Penikese Island opposite the town of New Bedford, on the south coast of Massachusetts. This was an unexpected gift from a wealthy New Yorker, who happened to own the small island. The station was mainly used for summer schools, where Agassiz's philosophy was taught to the younger generation. Although it did not long survive after Agassiz's death, because the endowment was insufficient to cover its running costs, it set the example for Woods Hole and other biological research stations founded later. Louis Agassiz died at home of a cerebral haemorrhage on December 14, 1873 at the age of sixty-six. Among his numerous awards and honours was the Copley Medal of the Royal Society. He was a man of such abundant energy and enthusiasm that many of his projects were left unfinished at his death, though the amount of work he accomplished was astonishing as the product of one man. His labours laid the foundation of scientific zoology in America and provided a stimulus to its pursuit all over the world.

CHARLES DARWIN (1809–1882)

The illustrious naturalist Charles Robert Darwin was the son of Robert Waring Darwin, an immensely successful and wealthy physician of Shrewsbury, and of Susannah, daughter of Josiah Wedgwood, whose pottery business was one of the early successes of the industrial revolution. He was the second son and the fifth of six children. Darwin could hardly remember his mother, who died when he was only eight so that he was raised instead by his elder sisters. He went as a boarder to Shrewsbury school, from which he derived little benefit, and thence to Edinburgh as a medical student from October 1825 until it became clear that medicine was not the career for him.

On October 15, 1827 he was admitted to Christ's College, Cambridge, with the intention of taking holy orders. The actual coursework at the university was of little interest to him but he made the acquaintance of some natural historians, notably John Stevens Henslow, the professor of botany, who encouraged him to become a full-time naturalist. Darwin was already planning a visit to the Canary Islands and hoped to persuade Henslow to accompany him. However this plan was superseded by an invitation he received, almost fortuitously, to sail round the world as naturalist on a voyage which lasted from the end of 1831 until 1836. Henslow helped to persuade Darwin's father to agree to him joining the expedition.

It was now over sixty years since Cook had taken Banks on his historic circumnavigation. Most parts of the known world were regularly visited by merchant vessels. The problems of preventing scurvy and determining longitude accurately had been overcome. Her Majesty's Ship *Beagle*, captained by Robert Fitzroy, was being sent to make a survey of the coastal waters of the southern part of South America. On a previous voyage Fitzroy had taken three inhabitants of the inhospitable island of Tierra del Fuego to England; now they were brought back to their homeland. The highlights of the voyage of the *Beagle* are described in detail in all the many biographies of Darwin. The ship called at the Cape Verde Islands, various places in South America, the Galápagos Islands, Tahiti, New Zealand, Australia, Mauritius and the Cape of Good Hope. In the course of his excursions, whenever the ship landed, Darwin made geological observations and collected both animal and plant specimens. He also collected fossils and the remains of animals that had only recently become extinct.

After his return to England, Darwin soon became an active member of the London scientific community. He first made his name as a geologist. A book that exercised a strong influence on him was Sir Charles Lyell's *Principles of Geology*, published in the early 1830s. Lyell, a close friend of the Hookers, was anti-scriptural but not an evolutionist. He particularly objected to the idea that human beings evolved from lower forms of animal life; for many people this was something they could not accept. Species died out and were replaced by other species, but the one did not change into the other. Variations occurred within natural limits but transmutation never. Lyell's book was built on examination of the fossil record. He was a uniformitarian, who believed in gradual change and natural causes, unlike the catastrophists. Darwin's demonstrations of the differences between cleavage and foliation of rocks, the relations of planes of cleavage to geological features over wide areas, of extensive elevations of land and their connection with earthquakes and volcanic eruptions, and his explanation of the formation of coral atolls were major contributions to science.

These have been overshadowed, however, by his even more momentous contributions to biology, as a result of his demonstration that evolution of living organisms has occurred, and of his discovery of the principle of natural selection of hereditable variation as the principal cause of evolution. From his original, orthodox, acceptance of the fixity of species, he was led gradually to a realisation of the truth of the origin of species by descent with modification, or evolution, by the results of his work in South America and the Galápagos archipelago, and the differences, related to the manner of life, between species of birds on the different geologically recent Galápagos islands. All these facts fell into place if it could be assumed that species were not originally fixed, but had arisen from previous species and had been modified during descent.

There has been a tendency to assume that the ideas which made him famous dawned on him on the famous voyage, rather on reflection after he returned to England. Even his diary gives this impression. He realised that it was useless to proclaim such a theory as evolution was a fact unless he also provided an answer to the question of how and by what mechanism it had occurred. The changes demonstrably made in cultivated plants and domestic animals by man since the neolithic period under the practice of artificial selection convinced him that selection must be the key to evolution of wild species in nature, but the problem was to discover the agent that performed the selection instead of man.

By 1838 Darwin had satisfied himself that if hereditable variation occurred, then individuals better adapted to their environment would leave more offspring and thus gradually change the type of species in the direction of more effective adaptation, but he did not yet know how in nature such a change was enforced. Then he read the essay on the principle of population by the pioneer demographer Thomas Robert Malthus (1766–1834) containing the (fallacious) argument that the geometrical rate of increase in the human population must outstrip the rate of increase of the food supply, with consequent hardship, misery and death of the poor. Darwin saw that this argument, translated to the case of plants and animals in nature, which cannot increase their food supply, explained the inevitable mortality that must hit the less efficiently adapted and favour the better adapted. His theory of natural selection of favourable hereditable variation, expelling inefficient variants from their ecological niches in the environment, and preserving and improving the favourable variants that replaced them in those niches, was then complete. He found that it explained otherwise inexplicable facts in the sciences of comparative anatomy, embryology, palaeontology and geographical distribution, to mention only the chief lines of enquiry.

Darwin published nothing on this subject for over twenty years, during which he completed the publication of his journal of the voyage of the *Beagle,* his geological researches and his long painstaking research on the species of living and fossil barnacles. Darwin was a semi-invalid for most of his life, a martyr to a medical condition that was never properly diagnosed. The symptoms began in the autumn of 1837, a year after he returned to England. For most of the years 1864 and 1865 he was a complete invalid and entirely housebound. Since his sickness grew worse as the battle over his work raged, it seems safe to say that stress was a factor in whatever it was.

In 1839 Darwin had married his first cousin Emma Wedgwood and continued to live in London for a few years but, partly because of his chronic illness, purchased and improved a property in the peaceful Kentish village of Downe, some twenty miles south-east of the city. Apart from short periods, they lived in Down House for the rest of his life. Darwin became very much a family man, with Emma superintending the running of the household in the typical Victorian manner. Since she was a sincerely religious person, it grieved her that Darwin had lost his faith long before. He rarely left home, but maintained an extensive correspondence with fellow-naturalists all over the world, one of whom was Alfred Wallace, who was in south-east Asia. As we shall see in his profile, Wallace had published a theory about species variation which Darwin thought contained 'nothing very new'. In writing to Wallace, amongst others, to request information about domestic animals bred for many generations in different parts of the globe he merely made polite reference to the article.

Members of Darwin's circle often came to stay at Down House for short periods. In mid April 1856 Lyell was there and Darwin outlined to him, for the first time, his theory of natural selection. Lyell saw that Darwin and Wallace were thinking on the same lines, and urged Darwin to get his theory into print without delay. Darwin said he was not yet ready to do so. Instead he began work on the book which eventually appeared as *The Origin of Species*. He began to write out his work on the origin of species, assembling the evidence he had collected over the years. One day, two years later, he was stunned to receive from Wallace on a remote island of the Malay archipelago an essay which contained a perfect summary of the work on which Darwin himself had been working for twenty-one years. Wallace's side of the story will be described in the next chapter but one. Faced with an ethical dilemma, Darwin consulted his friends Lyell and Hooker. It so happened that an emergency meeting of the Linnean Society was about to take place to decide how to fill the vacancy on the Council caused by the death of Robert Brown. Lyell and Hooker decided that Wallace's essay should be read at the meeting but that Darwin, to establish his claim, should prepare a paper for presentation at the same meeting. Although Darwin was preoccupied by family problems at the time – his infant son Charles had just died from scarlet fever, while both his wife and daughter were ill with diphtheria – he did what was required. Both papers were read by the secretary to a meeting of some thirty fellows. In a classic understatement the president of the society, in his annual report, remarked that 1858 was not marked by any of those striking discoveries which at once revolutionise the department of science on which they bear. Darwin at once started to revise his book, comparing Wallace's version of the theory with his own, and condensed it so as to expedite publication. The next year *On the Origin of Species* finally appeared. As Darwin expected it attracted a storm of controversy, since a belief in independent creation had been almost axiomatic even among biologists.

Although *On the Origin of Species* is Darwin's magnum opus he wrote other important books which contain further evidence for natural selection and much else, such as the operation of sexual election and other aspects of animal behaviour. In 1862 his *Fertilisation in Orchids* was published, in which he showed that plants have adaptations no less wonderful than those of animals, and in 1868 his *Variation of Animals and Plants under Domestication;* this was followed in 1871 by *The Descent of Man* and in 1872 by *The Expression of the Emotions in Man and Animals.* They were followed in 1875 by two more books which took botanists by surprise: *Insectivorous Plants,* which showed the extraordinary adaptations by means of which a plant like sundew captures, digests, and absorbs the substances of the bodies of flies, and *The Movement and Habits of Climbing Plants,* which not only analysed the processes by which plants climb, but also showed their importance as an adaptation whereby plants reach a height where their leaves are well exposed to sun and air, without the expenditure of time and synthesis of woody material required for the growth of the trunk of an independent tree. In 1876 Darwin published his *Cross and Self Fertilisation in the Vegetable Kingdom,* which contained the results of over ten years of experimental breeding and demonstrated the all-important fact that in the majority of cases, the products of cross-pollination are more numerous, larger, heavier, more vigorous and more fertile than products of self-pollination. This hybrid vigour is favoured by natural selection, and is explained by modern genetics.

In 1887 *Different Forms of Flowers on Plants of the Same Species* was published, revealing the astonishing dimorphisms and even trimorphisms of flowers by means of which cross-pollination is ensured. His *The Power of Movement in Plants*, published in 1880, describing the results of experiments of exposing roots and root-tips to light, was the starting point of the whole science of plant hormones and growth-promoting substances. Darwin's last work, *The Formation of Vegetable Mould through the Action of Worms* of 1881 measured the amount of soil sifted, ground to powder, and raised to the surface by worms in given areas in given times. Its importance is only now realised, after abuse of chemical fertilisers and pesticides has reduced the earthworm population to the detriments of fertility of soil.

Early in 1882 Darwin began to suffer heart pains. The night of April 18 he had a severe attack, and was faint and nauseous the following morning. He died at about four o'clock in the afternoon. For his work which places him on the same plane as Isaac Newton, Darwin received all possible honours from academies and foreign governments, but nothing from the British government except burial in Westminster Abbey, on April 26, 1882. By that time the theory of evolution had become almost universally accepted by scientists, but 'survival of the fittest' was open to various interpretations. Darwin's own views as to how evolution worked remained highly controversial, and his followers did not always agree with everything he wrote. Today, the implications of his work, confirmed by every branch of biology, continue to influence all fields of human endeavour.

The Darwin forbears were remarkable people: he was the grandson of Erasmus Darwin, a highly regarded physician with a wide range of interests that made him a leading figure of the Enlightenment. Darwin's sons also achieved some distinction. Of his four surviving sons two were scientists who became fellows of the Royal Society and were also knighted, one was not a scientist but in a varied career was president of the Royal Geographical Society, and one was a banker. The daughters all died in childhood or infancy, as did one other son who suffered from Down's syndrome.

6 From Gray to Galton

ASA GRAY (1810–1888)

Although Agassiz spent an important part of his career in the United States, it does not seem right to call him an American biologist, possibly Swiss–American. Audubon has a stronger claim, but the first native-born American biologist to be profiled here is the plant taxonomist Asa Gray. He was born to the farmer Moses Gray and his wife Roxana née Howard at Sauquoit, in Oneida County, New York State, on November 18, 1810. After finishing school he studied medicine at Fairfield Academy, where he developed an interest in botany. He practised medicine briefly before deciding on a career in botany. After a spell of schoolteaching he moved to New York City in 1834 where he studied under John Torrey in the chemical laboratory of the New York Medical School. Torrey, who was then the leading young botanist in the country, was generous in sharing his knowledge and contacts with Gray. The two men formed a lifelong friendship as a result of their common interest in botany, and collaborated on several projects, notably in writing *The Flora of North America* 1838–1843, the first attempt at a national flora, although only two volumes were completed. In 1835 Gray was appointed curator and librarian of the New York Lyceum of Natural History, where he wrote the first of a series of standard botanical textbooks.

The United States Navy began to send out voyages of exploration in the nineteenth century. In 1836 Gray was appointed botanist to the ill-fated Wilkes voyage to Antarctica, but because of delays in sailing resigned the following year. He made his first visit to European herbaria in 1838 in order to examine many American type specimens, and made contact with many distinguished European botanists, one of whom was Robert Brown. Afterwards he commented 'I have seen considerable of Brown, and like him much better than I thought, although he is certainly peculiar he has much more general information than I supposed, is full of gossip, and has a great deal of dry wit. He is growing old fast, and I suspect works very little now, and I fear there is not much more work now expected of him. He knows everything!' Brown expressed interest in an American tour, but he never crossed the Atlantic.

In 1842 Gray was appointed Fisher Professor of Natural History at Harvard University, thus becoming the first American to make a living from botany. He corresponded widely with both paid collectors and amateurs who sent him specimens and observations in return for help with identifications. As a result he was the first person who was able to obtain an overview of North American flora; his manual of the botany of the northern United States, first published in 1848, ran through several editions in his lifetime. He had married Jane Lathrop Loring in 1848; through her he was introduced to Boston society, which he never much cared for.

Gray was not a field botanist; he preferred to work quietly at his desk, analysing the information he received from his network of correspondents. For Harvard he created what was later called the Gray Herbarium and Library. Like Joseph Hooker, Gray was a pioneer plant geographer. When Hooker visited the United States in 1877 he was taken by Gray to see the western states. His most important discovery was of the similarity between the flora of eastern North America and the flora of East Asia, particularly Japan. His demonstration of this, which could hardly be the result of separate creation, was of great value to Darwin, who shared with Gray his ideas on evolution as they developed. Subsequently Gray championed

Darwin's work in the United States: 'no person understands my views and has defended them so well as A. Gray,' said Darwin, 'although he does not by any means go all the way with me'. Notably, Gray maintained that natural selection was compatible with Protestant theology.

Gray was not an American Huxley but largely through his efforts, within ten years of the publication of *On the Origin of Species* Darwinian evolution was accepted by nearly all working naturalists in America, despite the opposition of Agassiz. Although he preferred scholarship to politics, and was not a good lecturer, he successfully confronted his colleague Agassiz in debate. Gray was the founder of systematic botany in the United States, in recognition of which many institutions in America and Europe honoured him with awards and distinctions. He died in Cambridge on January 30, 1888, leaving no descendants.

SIR JOSEPH HOOKER (1817–1911)

As we have seen in Chapter Four, in a life devoted to the service of botany, Sir William Hooker began his career in East Anglia, where the Yarmouth banker Dawson Turner had persuaded him to invest part of his inheritance in a brewery, which operated in the Suffolk town of Halesworth. The Hooker family lived in a house at the brewery which was where Joseph Dalton Hooker was born on June 30, 1817. After attending Norwich Grammar School, like his father before him, he studied medicine at Glasgow University, after which he became assistant surgeon and naturalist on the naval expeditions to Antarctica 1839–1843, with the main objective of finding the south magnetic pole. His Majesty's Ships *Erebus* and *Terror* called at Madeira, on the outward voyage, and then St Helena, where a magnetic station was established. They went on via the Cape of Good Hope to Tasmania, from which they made their first reconnaissance of the Antarctic continent, prevented from landing by its great barrier of ice. There already was a magnetic station at Hobart but another was needed on the Bay of Islands so they sailed to New Zealand, from which they made another probe into Antarctica. Both vessels were severely damaged in a storm so it was decided to spend the winter in the Falklands where they could be repaired. In November 1843 they rounded Cape Horn, reached the Cape of Good Hope in April the next year and were back in England by September. Their observations had achieved the primary objective of locating the south magnetic pole with some precision. In addition Hooker brought home a huge collection of botanical and other specimens, many of which proved to be new to science. His subsequent botanical narrative of the expedition *Flora Antarctica* 1839–1843, *Flora Novae Zelandiae* 1853–1855 and *Flora*

Tasmaniae 1855–1860 placed him in the forefront of taxonomic botanists. He gave cautious approval to the theory of natural selection in his celebrated introductory essay to the last of these, but soon became a vigorous advocate of Darwin's work.

In 1847 he departed for India where he spent three years collecting in Bengal, Sikkim, eastern Nepal and Assam. His Himalayan journals of 1854 closely observe the fauna and ethnography as well as the flora of the region, and later travellers testified to the accuracy of his surveying. *The Rhododendrons of the Sikkim–Himalaya* 1849 appeared while he was still out there, and his love for India culminated in his seven-volume *Flora of British India* 1872–1897. Further expeditions that he later undertook were to Syria and Palestine in 1860, the Atlas Mountains of Morocco in 1871 and the western United States in 1877.

On his return from India in 1851 the younger Hooker was elected to the Royal Society at a crowded meeting of his supporters. He was beginning to be recognised as the greatest botanist of his time, the greatest naturalist apart from Darwin. After he first met Darwin in 1839, they became life-long friends. Although he was well aware of Darwin's researches, they both worked independently. At the famous British Association meeting in Oxford in 1867 Henslow, a friend of the family, had taken along his clever daughter Frances Harriet, and they had become engaged. Now, after being separated for almost four years, they could proceed to be married. She was both by birth and training able to help in his work and share all his scientific aims and enthusiasms. They began married life rather dependent on their parents for money, until in 1855 he was appointed assistant director of the Royal Botanic Gardens at Kew. Ten years later he succeeded his father as director. The sudden death of Frances in 1874 left him with six children to look after. He lost no time in quietly marrying again, this time a widow named Hyacinth, Lady Jardine, whom he first met at the British Association in Bath in 1864. Her late husband, a baronet, was a distinguished amateur naturalist. She performed the difficult role of stepmother quite admirably.

The younger Hooker continued his father's work by developing Kew as an international centre of scientific research, but he did not neglect its role in economic botany. We owe to the plant collectors of Victorian times many of the plants that grow in our gardens today, but plants of economic importance were also collected. Rubber-producing trees of Brazil were introduced into Ceylon and Malaya where plantation rubber became a major industry; against this competition Brazilian exports of wild rubber collapsed. Little effort was made to prevent other nations exploiting economically the scientific work being done at Kew; thus German interests benefited greatly from the sisal plantations of East Africa at the expense of the Mexican sisal industry. In France Geoffroy's son Isidore was instrumental in founding the Société impériale d'acclimatation which aimed at identifying animals that could be moved successfully from their natural habitat to another part of the world, an enterprise which was copied in England, but without the success of the botanical transplantations.

Under Joseph Hooker's directorship botanists were sent from Kew to all parts of the world to collect material and obtain information. Botanical gardens, linked to Kew, were established in Singapore, Melbourne and other places in the British Empire. At Kew itself the Jodrell Laboratory was opened for investigations into plant anatomy and physiology in 1876, the Marianne North gallery in 1882 and the rock garden in 1882. The Kew Herbarium is still arranged according to a plant classification devised by Hooker and his colleague George Bentham, and published as the three-volume *Genera Plantarum* in 1862–1883. Among the many books and articles he produced a monograph of the 'life and labours' of his much-admired father. He was president of the Royal Society from 1873 to 1878. In 1885, weighed down by official duties he was expected to perform without any assistance, he resigned as director of the Royal Botanical Gardens, after thirty years in office. In planning for his retirement he had bought a plot of land near Windsor Great Park, and had a house built which was ready for occupation in 1882. He had twenty-six years of active and useful life before him, writing up the fruits of his research, and receiving honours from home and abroad, too numerous to mention. Joseph Hooker was knighted, like his father, and received into the Order of Merit (an honour limited to 24 members of the arts and sciences) in 1907. He died on December 10, 1911.

RICHARD SPRUCE (1817–1893)

Richard Spruce, one of the unsung heroes of plant-collecting, was born on September 10, 1817 at Ganthorpe, near Castle Howard in Yorkshire, where his father was village schoolmaster. His mother, said to be 'of exceedingly nervous temperament', was related to the painter William Etty (1787–1849); when she died in 1829 his father remarried. There were eight daughters from this second marriage as well as three from the first. Spruce, the only son, was educated by his father, it seems, and followed him into the teaching profession. From an early age he was interested in natural history. Between 1839 and 1844 he was a schoolmaster, teaching mathematics in the Collegiate School at York while devoting his spare time to botanical studies. He made field trips to localities throughout Yorkshire, especially Teesdale. In the summer of 1844 he visited the west of Ireland, where he stayed with Thomas Glanville Taylor (1786–1848), one of the world's leading authorities. Although handicapped by illness he inspected numerous botanical sites in Kerry and Cork. Already he was building up a sizeable herbarium and publishing papers which brought him to the attention of Sir William Hooker and other eminent botanists. He specialised in mosses and liverworts, publishing his first paper, on those of Eskdale, in 1841, and subsequently published papers on those of Teesdale and of other parts of Yorkshire. Spruce was recommended to Hooker for a twelve-month botanical expedition to the Pyrenees, helping to defray his expenses by the sale of the flora he collected. In addition to its botanical value the expedition also helped to improve his health. He described the results in three letters published in Hooker's *London Journal of Botany*. Two more works followed: *Notes on the Botany of the Pyrenees* and a more elaborate article *The Mucci and Hepaticae of the Pyrenees*. This work was interrupted on several occasions when he had to deputise for his ailing father at the village school.

Already Spruce was hoping for a chance to visit South America and preparing for it by studying tropical plants at Kew. The opportunity came when Hooker chose him to undertake a botanical expedition to the Amazon Valley and the Andes. As an emissary of the Royal Botanic Gardens his mission was not to theorise but to acquire an intimate knowledge of the flora. The British Museum and the Royal Botanic Gardens were to have first choice of the material he sent back; the rest was to be sold for him by an agent. He sailed for the city of Para at the mouth of the Amazon, now known as Belem, in June 1849, at the age of thirty-one. After a few months he travelled up river to Santarem, at the mouth of the Tapajos, where he met the zoologist Alfred Russel Wallace and the lepidopterist Henry Walter Bates, of whom more later. Spruce explored the river Trombetas to a point close to the borders of British Guiana. Towards the end of 1850 he reached Barra, now Manaos, where the river Negro meets the Amazon. The following year he explored the forests around Barra after which he set off with six companions for the upper reaches of the Rio Negro, where he spent the next three years. During that time he crossed over to the Orinoco and penetrated some distance into Venezuela. Spruce discovered many plants new to science in this region, including 200 species of fungi in the rainforest of the little-known Uaupes river.

Spruce returned to Barra at the close of 1854, and then ascended the Amazon by steamer to Nantua in Peru, proceeding by canoe to Tarapoto on the eastern side of the Andes, where he stayed two years and collected, within a twenty-five-mile radius, 250 species of ferns. In 1857 he again descended the Amazon and went up the Pastasa to Ecuador, reaching Banos. Six months later he moved on to Ambato, which he made his headquarters for two years and from where, despite hostilities then in progress, he explored the Quintensian Andes. Among his discoveries were a number of plants with medicinal properties, including the datura and coca plants. It was his samples of the latter, sent to Berlin, which helped Niemann isolate the active principle of the alkaloid cocaine.

In 1859 he was commissioned by the India Office to collect seeds and young plants of the cinchona, the source of quinine; he managed to obtain 1 000 000 seeds and 600 plants on the western slope of Chimporazo, which were dispatched to England through Guayaquil. Seedlings and cuttings from the cinchona, collected by Spruce, were surreptitiously brought from Ecuador, grown at Kew, and were sent to India where millions benefited from an inexpensive treatment of malaria, although British residents were first priority. Spruce's report on this expedition was published in 1861. During this period on the altiplano his health, never good, seriously deteriorated. Spruce remained on the Pacific coast for several years in the hope of recovering, but meanwhile lost the whole of his modest savings in the failure of an Ecuadorian bank; he returned to collecting for three years, finally embarking for England in 1867 after an absence of eighteen years.

Spruce's collections were not confined to botany. During his travels he made notes of the economic and medicinal uses the Indians made of the plant-life around them. He learnt and recorded vocabularies for twenty-one Amazonian languages, made hundreds of drawings, and mapped three previously unexplored rivers. The 7000 flowering plant species he collected were catalogued, as well as numerous ferns, mosses, lichens and fungi. Spruce was awarded an honorary Ph.D. degree from Dresden, was elected a fellow of the Botanical Society of Edinburgh and an honorary fellow of the Royal Geographical Society, with similar honours from other bodies.

Spruce, who never married, retired to the tiny hamlet of Coneysthorpe on the Castle Howard estate, near to his birthplace. Obliged to live frugally on a meagre £100 pension from the British and Indian governments he spent the remaining seventeen years of his life on sorting his collections, forced by ill health to do much of this work lying down. He also maintained a lively correspondence with fellow botanists throughout the world and wrote numerous scientific papers. He died at Coneysthorpe on December 28, 1893 at the age of seventy-six and was buried in the churchyard of Terrington nearby. After his death his notebooks were edited by Wallace and published in 1908 as *Notes of a Botanist on the Amazon and Andes*. This work reflects the author's gift for analysis, description and narrative, along with his dry humour. His name was commemorated by the moss *Sprucea* and the liverwort *Sprucella*. Too self-effacing and modest about his achievements, he never aspired to be one of the scientific elite.

SIR FRANCIS GALTON (1822–1911)

The Galtons were a Quaker family and continued to be so although the pacifist Society of Friends disapproved of their gun-making business. Because of this the paternal grandfather of the subject of this profile, who had married into the Barclay banking family, founded the Galton bank, in partnership with his son Samuel Tertius Galton. The latter married into the Darwin family and was to be the father of the future scientist. Erasmus Darwin had five children by his first wife, including Robert, the father of Charles Darwin, and then seven children by his second as well as two by his mistress. It was Violetta, one of the daughters of Erasmus by his second wife, who married Samuel and gave birth to Francis.

He was educated at home until he was five, then at various schools until he left King Edward's School Birmingham at the age of sixteen. His mother was keen for her cleverest son to enter the medical profession, following the example of Erasmus, and so he spent one year gaining practical experience at Birmingham General Hospital before entering the medical school at King's College London. He took the advice of his half first-cousin, the famous naturalist, to interrupt his medical education and read mathematics at Cambridge. He took full advantage of the social opportunities that the university had to offer and made many friends who were useful to him in later life, but following the first of a series of nervous breakdowns he graduated with only a pass degree.

In 1844 his father died, leaving him financially independent. Of his father Francis wrote that:

> he also had scientific interests, like Erasmus, but he became a careful man of business, on whose shoulders the responsible work of the bank chiefly rested in times of trouble. Its duties had much cramped the joy and aspirations of his early youth and manhood, and narrowed the opportunity he always eagerly desired, of abundant leisure for

systematic study. As one result of this drawback to his own development, he was earnestly desirous of giving me every opportunity of being educated that seemed feasible and right.

Be that as it may, Francis dropped his medical studies and started to satisfy a strong urge to travel and sow his wild oats. Before going up to Cambridge he had been to parts of Eastern Europe that were not usually visited. Now he ventured further afield, making an extended tour of Egypt and Syria. However his most adventurous trip was a two-year reconnaissance of south-west Africa, nowadays called Namibia. This involved a journey of some 1700 miles, much of it through unexplored country, on which he made an accurate determination of the latitude and longitude of the places he passed through. Once he was back in England he was awarded a gold medal by the Royal Geographical Society, for having explored a part of Africa previously unknown. He also wrote a book about it called *On the Art of Travel, or Shifts and Contrivances Available in Wild Countries*. This proved very useful and popular, running to eight editions.

Being now over thirty he was ready for marriage. His wife, Louisa Butler, came of an intellectually distinguished family and was herself an intelligent woman but she did not have enough to occupy her mind.

She was not interested in domestic affairs and they had no children. She played no part in his scientific life, as far as is known; her interests were more in the fine arts. When they went away it was always to one of the fashionable spa resorts of Europe. She tended to exaggerate her ill health, so that he kept her company, but then he became unwell, because he was unable to continue working.

After a few years in rented accommodation they purchased a house in South Kensington, where he lived for the rest of his life. This was convenient for the Royal Geographical Society of which he became a leading member. His opponents described him as a man of eccentric genius and tastes who inflicted his corvine voice on every meeting of the Society. He made a practice of counting the number of fidgets as an index of the boredom of the audience. As well as the Royal Geographical Galton was active in other scientific bodies, notably the British Association for the Advancement of Science, to which he contributed many papers over the years, on a wide range of topics. He was in his element as a member of the scientific elite who met each other at the Athenaeum. Socially and politically conservative, he had little sympathy for liberal causes.

In her diaries Beatrice Webb wrote this about Galton:

> Even today I can conjure up, from memory's misty deep, that tall figure with its attitude of perfect physical and mental poise – the clean-shaven face, the thin, compressed mouth with its enigmatical smile; the long upper lip and the firm chin, and, as if presiding over the whole personality of the man, the prominent dark eyebrows from beneath which gleamed, with penetrating humour, contemplative gray eyes.

One of his colleagues described him as follows:

> He had intelligent rather eager blue eyes and heavy brows, a long straight mouth, and a bald head. Without an atom of vanity he held to his own opinions and aims tenaciously. He was very clever and perfectly straight in all his dealings with a strong sense of duty, but his mind was mathematical and statistical with little or no imagination. He was essentially a doctrinaire, not endowed with much empathy and not adapted to lead or influence others. He had no tact and was unable to make any allowance for the failings of others.

These personality characteristics are strongly suggestive of Asperger's syndrome and yet there are contra-indications. One of his great-nieces wrote that if asked her opinion she would say 'what a pet he was and how good-tempered and full of delightful naïve sayings, and that everyone wanted to kiss him.'

The publication of Darwin's *On the Origin of Species* marked a turning point in Galton's intellectual life. He had little interest in natural history or in the evolution of plants and animals but he focused on the implications of Darwin's theory for humans. This was the source of Galton's interest in heredity and the possible improvement of the human race. Eventually this led him into deep waters but his book *Hereditary Genius* created a considerable stir at the time it first appeared in 1869 and initiated an ongoing debate as to whether certain types of mental ability were inherited or the result of upbringing (he regretted that he had not called his book *Hereditary Ability*). After examining various lists of professional people he concluded that 'intellectual capacity is so largely transmitted by descent that, out of every hundred sons of men distinguished in the open professions, no less than eight are found to have rivalled their fathers in eminence'. He showed that this was still true when he included a wider range of professional people, including statesmen, military commanders, literary men, scientists, poets, musicians, painters and divines. Darwin wrote to say he did not think he had ever in all his life read anything more interesting and original. Galton started a debate over what was called nature versus nurture which is still active today.

Galton's next book, his *Inquiries into Human Faculty and Development* of 1883, incorporates the results of his original investigations into a wide variety of interesting questions. For example he was interested in composite portraiture. When he tried this out by superimposing photographs of a number of Jewish boys he found this produced what he regarded as a typical Jewish face. He wondered if the same process might be a useful guide to criminal behaviour and tried it out on prisoners who had been convicted of different types of crime; this was not a success. He pioneered the use of identical twins for scientific research; these are usually reared in the same environment but afterwards live apart from each other. He obtained some interesting results from the thirty-five sets of such twins he studied; today research on twins continues to provide insight into the question of nature versus nurture (see Segal 2006). He also investigated the use of fingerprinting as a means of identification, after establishing that the probability of two individuals having the same prints was negligible. Galton played an important role in putting the use of fingerprints as evidence onto a firm scientific footing. His system of classifying fingerprints was over-elaborate but it led to the practical one which is still in use today. He was also interested in mental imagery, specifically in the 'number forms' which some people experience; such phenomena are still only poorly understood.

Galton's work was generally well received in scientific circles but sometimes he succeeded in scandalising Victorian sensibilities. For example he published a note about the efficacy of prayer, arguing that if prayer was effective this should be reflected in the lifespan of members of the royal family, who were much prayed for, or the clergy, who did much praying. No such effect was observable in either case. Also he argued that the enforced celibacy of intelligent members of the Catholic priesthood deprived the world of the intelligent children they might have had if they had been able to marry. Florence Nightingale shared his enthusiasm for social statistics and suggested to him that a professorship or readership in this field would be desirable at Oxford. Galton believed that London would be a better place for it: such a person would be too isolated at Oxford, or Cambridge for that matter, and he endowed such a post after his death.

The science of the hereditary improvement of the human race by selective breeding, rather than leaving it to natural selection, is known as eugenics. Galton coined this term in 1865 and advocated the idea for the rest of his life. He promoted both positive eugenics, such as the encouragement of early marriage between talented men and women, and negative eugenics, the discouragement in the case of those considered unfit. Galton realised that research would be necessary before eugenics could be put into practice, and so in 1884 established an Anthropometric Laboratory at University College London, which was later to be renamed the Galton Laboratory, and devoted to biometrics. He also endowed a research fellowship in eugenics and when he died he left the university funds to endow a chair in the subject at the college. The latter was held first by Karl Pearson and then, as we shall see, by Ronald Fisher, who were in favour of positive eugenics, although the next holder was Lionel Penrose who was not.

The last decade of Galton's life was spent expounding eugenics as a social, political and religious creed. In Britain public opinion was generally against the idea of negative eugenics but not in the United States, where by 1914 legislation had been passed in thirty states preventing the marriage of the 'mentally deficient'. In Nazi Germany the sterilisation of those suffering from eight allegedly hereditary disorders was made compulsory, and it is estimated that about 400 000 sterilisations were carried out. Later sterilisation was replaced by euthanasia as being more cost-effective. Today, although the idea of negative eugenics is discredited, the idea of positive eugenics is by no means dead.

Someone who met Galton first as an octogenarian recorded that 'he seemed quite short, an impression accentuated by his stooped walk. His deep blue eyes were surmounted by prominent brows set on either side of a well-chiselled nose. He appeared simple and unaffected and spoke in a soft voice with a smooth, sweet quality. He was unfailingly courteous and brought out the best in everyone by gently probing for their chief interests and making these the topic of conversation.' 'I could never find a subject that Galton was not willing and eager to discuss', this person added, 'and he always managed to throw new light on matters upon which one liked to believe oneself an expert.'

Galton died suddenly of heart failure on January 17, 1911, at the age of eighty-eight, his wife having predeceased him by some fourteen years. A few months earlier he had learnt that he was to be awarded the prestigious Copley Medal of the Royal Society. Among many other honours he was knighted in 1909. Although Galton's name is linked especially with genetics, he was a man of diverse interests and many achievements. Those already mentioned do not complete the list by any means; for example, he was interested in meteorology, where the concept of anticyclone was due to Galton. As Maynard Keynes put it:

> there was no-one who has possessed in purer essence than he the spirit of universal scientific curiosity ... it was not the business of his particular kind of brain to push anything very far. His original genius was superior to his intellect, but his intellect was always just sufficient to keep him just on the right side of eccentricity.

Beatrice Webb was fascinated by his continuous curiosity about, and rapid comprehension of, individual facts, whether common or uncommon, his faculty for ingenious trains of reasoning, and his capacity for correcting and verifying his own hypotheses, by the statistical handling of masses of data, whether collected by himself or by other students of the problem.

7 From Mendel to Potter

GREGOR MENDEL (1822–1884)

Johann Mendel was born July 22, 1822 in the Moravian village of Heizendorf, close to where the borders of Germany, Poland and the Czech Republic meet; this is now the Czech village of Hyncice. He was the second child of Anton and Rosine Mendel. They also had two daughters, Veronica who was older than Johann and took after her father, and Theresia who was younger and took after her mother. Anton was a farmer, with a special interest in fruit-growing, an interest which he passed on to his son. Many of Rosine's forebears, including her father, were professional gardeners employed at the local manor.

At the village school the children were taught natural history and natural science, as well as the ordinary subjects of elementary education. Because he already showed unusual ability Johann was sent to a higher school at Leipnik (now Lipnik), about thirteen miles away, where he soon distinguished himself, and then to the high school at Troppau (now Opava), twenty miles away. The family were not at all well off; it was a struggle to find the school fees, especially when Anton had to give up farming after an accident at work, and to avoid interrupting Johann's education, transferred the family farm to Veronica's husband. Meanwhile Johann had enrolled as a student in philosophy at an institute which was affiliated to the university of Olmutz (now Olomouc), trying unsuccessfully to support himself by private tutoring. When his health began to suffer he returned home for a while and then his younger sister sacrificed part of her dowry to enable him to complete the two-year philosophy course, which included mathematics and physics. What he learned about the scientific method, especially the design of experiments, proved of great value in his future work.

He left the Olmutz institute with a strong letter of recommendation from his professor of physics to the Altbrunn monastery at Brunn, where the professor had spent twenty years himself. These religious foundations were important centres of learning and scientific research. So the next step in Mendel's career was a move to the Moravian capital, now the Czech city of Brno, where he was accepted as a novice by the Augustinian monastery, and

given the name of Gregor in October 1843. This freed him from financial worries and provided him with an environment in which he could pursue his scientific interests. Almost all the monks were engaged in educational, scientific or artistic activities of one kind or another. Since the foundation's income came mainly from its estates it followed that the improvement of agriculture was important. As a result of this one of the monks, an able botanist, had built up a huge herbarium at the monastery, including a complete collection of the Moravian flora. Since the monk in question had recently died Mendel at once took an interest in the herbarium and worked there when he was not occupied with the studies prescribed for his probationary year.

After Mendel had completed two years at the monastery he was sent on a course of four years' theological study at Brunn theological college, during which he took the vows of the Augustinian order, and was ordained to the priesthood at the age of twenty-five. The last year of the course was 1848, the famous year of revolutions. The monastery was unaffected but Mendel's parents benefited from the abolition of the córvée, the obligation to work part-time for the lord of the manor in which they lived. After an unsuccessful trial period as a parish priest Mendel began temporary work as a schoolteacher at the high school of Znaim, a picturesque old town forty-eight miles north-west of Vienna. When this came to an end he was appointed to a similar position at the Brunn Technical School. Meanwhile Mendel had been persuaded to take the examination for high school teachers, in natural history for the entire curriculum and in physics for the lower school. The examiners were professors from the University of Vienna. All but one was satisfied with his performance, but he failed to pass.

Mendel had received no relevant university study, but was self-taught; one of the examiners helped to arrange for Mendel to study at the University of Vienna. He attended lectures on experimental and mathematical physics, also chemistry, botany and zoology. He also attended some private lectures on entomology and became a member of the Zoological and Botanical Society of Vienna. Mendel left the university and returned to Brunn in 1853. A modern school had recently been founded there where the traditional teaching of the humanities was supplemented by some lessons in science. Mendel obtained another temporary position there, teaching physics and natural history to the lower school. His years as a schoolmaster were good years; his pupils had a high regard for him. He proved himself an exceptionally good teacher but when he retook the examination he had failed earlier he was again unsuccessful, and he never became a fully qualified teacher.

In Brunn there were numerous artistic, literary and scientific societies, of the amateur type, but in 1862 some young and enthusiastic citizens founded a society for the study of natural science, in which Mendel played an active part from the start. The leading light of the society was the versatile Gustav von Niessl, professor of geodesy at the Brunn Technical School, who was interested in astronomy, botany and meteorology as well as geodesy, and who was also chairman of the Brunn Musical Society. Mendel often used to discuss the problem of the origin of species with Niessl, and owed a great deal to his insight and clear thinking. Mendel kept trying to produce permanent variations by transporting plants from their natural habitat, but without success. He concluded that nature does not modify species in that way. Mendel was far from being an adversary of the Darwinian theory but thought that something was lacking in it. Among the many scientific books he collected he bought Darwin's books as soon as they appeared, and annotated his copies. There is no evidence that Mendel ever got in touch with Darwin or that Darwin was aware of Mendel's work. If he had been, the history of evolutionary biology would have been different.

Mendel owned a microscope which he used for botanical work, although he found it extremely trying to the eyes. He also used to breed mice in his rooms but his tastes were more botanical than zoological. Also his superiors might have frowned on breeding experiments with animals. In 1856 Mendel, at the age of thirty-four, began his experiments in the crossing of the edible pea, and seven years later, when he was forty-one, he had substantially finished them. He was allocated a strip of the monastery garden, a hundred and twenty feet by a little over twenty; his letters refer again and again to the inadequate space he had for his researches. When he was elected abbot and prelate in 1868 he was no longer confined to this small strip for his experiments but now was hampered by lack of time rather than lack of space, so that effectively his scientific work slowed to a halt. There was no response from the scientific world to his discoveries, reported in his classic monograph *Versuche über Pflanzenhybriden*. This was published in the proceedings of the Brunn Society for the Study of Natural Science, exchanged with more than 120 other societies, universities and academies at home and abroad. He also sent reprints to some forty biologists who he thought might be interested in his work; this led to some discouraging criticism from the Swiss botanist Carl Nageli (1817–1891).

During the last decade of his life Mendel used his spare time to work on apiculture and meteorology, especially the latter. He had given up his schoolteaching with regret when he became prelate. He travelled a good deal, was certainly in Rome and probably in England. A dispute with the authorities over the taxation of his monastery, one of the wealthiest in the country, caused him a great deal of trouble, and he grew more and more embittered. He suffered from chronic kidney disease and heart trouble; in addition he was corpulent and a heavy smoker. Mendel died on January 6, 1884. In 1900 Hugo de Vries in Amsterdam, Carl Correns in Tübingen and Erich von Tschermak in Vienna, working independently, reported on research which confirmed Mendel's, of which they had only just become aware. The wider recognition of Mendel's law was finally achieved on the appearance in 1909 of *Mendel's Principles of Heredity* by William Bateson, who gave the name 'genetics' to the study of heredity and its variation. Mendel's experimental work, designed after long contemplation of the problem, painstakingly executed on an extensive scale, intelligently analysed and interpreted, and presented straightforwardly and clearly, yielded results of such general and far-reaching significance that his paper became the basis for the science of genetics.

ALFRED WALLACE (1823–1913)

Alfred Russel Wallace, who appeared in the profile of Darwin, was born on January 8, 1823 in a remote and picturesque market town in Monmouthshire. He was the eighth child of Thomas Vere Wallace and Mary

Anne née Greenell. His mother's forebears were French Huguenots, who had fled to England at the end of the sixteenth century, taking the name of Greenell. His father's forebears were middle-class Anglicans. Thomas Wallace trained as a lawyer but never practised. He squandered the family's financial resources through a sequence of ill-judged business dealings. He moved his family from Hertford, north of London, to Usk about five years before Alfred was born. They occupied a spacious riverside cottage, where they could live much more economically. However, in 1828 his mother was left some money which enabled the family to return to Hertford, where Alfred received his only formal education at the local grammar school. Otherwise he was taught by his father and older brothers, especially John, six years his senior, who was a talented mechanic. In 1832 Alfred's twenty-two-year-old sister Eliza died of tuberculosis; three of her sisters had died previously. Then his father lost the last of his savings through poor investment in real estate, while the executor of the Greenell estate, of which Anne was a beneficiary, revealed that he had used her inheritance, quite improperly, to try and settle his debts. His bankruptcy left the Wallace family destitute; the parents moved to a small cottage in the village of Hoddesdon, near Hertford, where their remaining daughter Fanny was employed as a governess. Alfred was withdrawn from school and sent to join his brother John, by this time an apprentice carpenter in London. He was not expected to work; instead John encouraged his brother to attend the meetings of the London Mechanics Institute, where he was exposed to radical ideas.

However this was only a temporary arrangement. In the summer of 1837 Alfred joined his oldest brother William as an apprentice land-surveyor. William was twenty-seven, intelligent, well-read and worldly. Despite their fourteen-year difference in age the brothers got on well in the next seven years. They settled in the Welsh town of Neath, off the Bristol channel. Due to the passing of the Tithe Commutation Act a complete survey of the English countryside was needed and William earned a living from this work. Encouraged by William, Alfred developed an interest in natural history and had a good grasp of the basic principles of geology. Thomas Wallace died in 1843, at the age of seventy-two, leaving his dependents without financial support. His widow was obliged to work as a housekeeper; Fanny, still unmarried at thirty-one, emigrated to the United States, where she was employed as a teacher at a small college in Macon, Georgia. The youngest son, Herbert, was withdrawn from school and apprenticed in London. The rest of the family had jobs, but it was a time of widespread unemployment. When William found himself out of work, Alfred went back to London in search of a job. Luckily he obtained one as a schoolteacher in Leicester, teaching drawing, surveying and mapping to schoolboys. There he was fortunate to meet the lepidopterist Henry Walter Bates (1815–1892), then just nineteen years old. They went off together at weekends collecting insects in the surrounding countryside.

In 1843 William died suddenly from pneumonia at the age of thirty-six. His surveying business was heavily in debt. With John's help Alfred tried to revive it but without success. Again luck was on his side. The great railway boom had just started and there was well-paid work for surveyors when companies sought to obtain parliamentary approval for their plans. In fact less than a tenth of the planned mileage was ever constructed and speculators often lost their investment. Next Alfred and John went into partnership to set up an architectural, engineering and building firm. They bought a small cottage near Neath in which they lived with their mother. Alfred kept in touch with Henry Bates and gave lectures at the Neath Mechanics Institute, founded by a friend of William's, which the firm designed and built. It was his first experience of lecturing but he made a good impression on his working-class audience.

This was the period when Robert Chambers set out a materialistic explanation for the creation of the universe and of life, rejecting a literal interpretation of the Judaeo-Christian conception as put forth in the Bible. The whole of our firmament, wrote Chambers, was originally a diffused mass of nebulous matter filling the space it still occupies. Gradually the stars, our Sun and the planets formed. At first the Earth was too hot to support life. Organic life began after the appearance of dry land, but the Earth had to go though some changes before conditions were fit for the emergence of terrestrial life. The Almighty could not have brought forth each individual species but rather through the introduction of natural laws which were expressions of his will. Like most of his contemporaries Chambers envisioned a great chain of being. Chambers' contentious book, called *Vestiges of the Natural History of Creation*, immediately came under attack from all sides. Anticipating this Chambers published it anonymously. The religious critics dismissed anything that questioned the account given in Genesis. The majority of scientists also dismissed the book as outlandish; Darwin and Huxley thought it amateurish. However, the general population took it seriously.

Wallace was electrified by Chambers' book; it altered his perception of the natural world, but Bates was unimpressed. In 1847 Alfred's sister Fanny, who had just returned from the United States, took both Alfred and John to Paris for a week, during which they toured the extensive collection of specimens housed at the Muséum d'histoire naturelle. Alfred was deeply impressed by the enormous variety of species of insects. Back in London he visited the natural history section of the British Museum and was even more impressed by the enormous number of entomological specimens, from all parts of the known world. Wallace persuaded Bates that they ought to try and go to some relatively unknown area and focus their attention on one particular family of insects. Hopefully they might be able, through the facts they gathered, to solve the mystery of mysteries, the origin of species. Neither of them was wealthy, like Alexander von Humboldt, or could obtain government sponsorship, like Darwin or Banks, nor were they medically qualified, able to serve as medical officers on a voyage of the Royal Navy. They hoped to cover the expenses of their expedition by selling specimens to wealthy collectors and institutions.

There were good reasons for selecting Brazil for their expedition. Few biologists had yet penetrated to the heart of Amazonia, but Bates and Wallace were not the first. Two Germans, Johannes von Spix and Carl Friedrich von Martius, had ascended the great river from Para in 1819–1820. Between 1822 and 1844 no foreign explorers had been admitted to the country, but now conditions were normal again. The vast, tropical rainforest, with its rich biodiversity, excited great interest in the London scientific community and was the perfect place for two neophytes to make names for themselves. So in 1848 Bates and Wallace started to make arrangements for an expedition to Amazonia. They met in London to study the treasure trove of specimens of South American fauna in the British Museum, noting gaps that they might hope to fill. They also made the necessary pilgrimage to Kew, where the huge conservatories housed live plants while dried specimens were exhibited indoors.

Wallace had studied Humboldt's *Personal Narrative of a Journey to the Equinoxial Regions of the New Continent* and Darwin's *Voyage of the Beagle*, but the book that seemed to fire his imagination was the little-known *A Voyage up the River Amazon, Including a Residence at Para*, by the respected American amateur naturalist William Henry Edwards. By chance Edwards was in London at this time; he gave them practical advice and letters of introduction to members of the expatriate community in Para. Others gave them advice about the care and preparation of specimens. They made an agreement with Samuel Stevens, an honest and reliable agent, to sell the materials they would be sending him, explaining how to ensure that they arrived in good condition. Hooker wrote to the British Foreign Office in support of their applications for passports and to the Brazilian authorities to confirm their identities and scientific bona fides.

Before leaving Wallace wound up the architectural business he had set up with his brother John, who was left to support himself, their mother and sister Fanny as best he could. In the spring of 1849 John decided to give it up and sail for California during the mad rush for gold. Fortunately Fanny married that year and moved to London with her mother and husband, a professional photographer. Alfred's remaining sibling Herbert remained in his job at the iron works in Neath. The two naturalists were lent some money by Bates' father, but otherwise they only had their personal savings to fund their expedition.

On April 26 they boarded the merchant vessel *Mischief* at Liverpool, and a month later crossed the equator and anchored off the small village of Salinas, the only port of entry to the vast Amazonian watershed, waiting to be piloted though the intricate channels to Para. They rented a house in the rainforest and began scientific work. After a slow start, by the end of two months they could send Stevens 3635 insects in addition to twelve chests full of native plants. They chartered a large canoe, captained by an American expatriate with local knowledge, in which they sailed up the Tocantins river, the third largest in the Amazon system, and found plenty of new flora and fauna. On the way back Wallace was incapacitated by a gunshot wound when his gun misfired – he was a poor shot.

After only four months, however, the Bates–Wallace partnership ended in acrimony, through some serious disagreement about some matter unknown; the two naturalists avoided each other for the whole of the next year. During this year in Para Wallace had become adept at shooting, skinning and preserving his specimens, and fluent in the lingua franca of the country, but without Bates he lacked a trusted English-speaking companion. So he made arrangements for his younger brother, twenty-one-year-old Herbert, who had been somewhat adrift for the past seven years, to join him in Brazil. Eager or not, Herbert, who now preferred to be called by his middle name of Edward, agreed and arrived in Para in June 1849.

The next month the brothers were ready to set out for the heart of Amazonia. Wallace barely knew his travelling companion, who was a self-styled intellectual, with a love of poetry and distaste for manual labour. They set forth in a small sailing vessel and after ten days were on the Amazon itself, awed by its immensity. Twenty-eight days after leaving Para they arrived at Santarem, their destination. Wildlife was plentiful on both sides of the river but Wallace was struck by the difference between the two sides.

They continued to Barra where they found Bates, and the two naturalists made up their quarrel, whatever it may have been. In late March they parted company again, this time amicably, after agreeing to explore different parts of the Amazon basin. They did not meet again for another twelve years. Edward was not suited to the life of an explorer and so Alfred left his brother behind in Barra to return to England while he went up the Rio Negro for the next five months, reaching the town of San Carlos, on the Venezuelan border, where Humboldt had turned back.

He continued on through Venezuelan territory, an arduous journey on foot, until he reached the headwaters of the Orinoco and a whole new flora and fauna. In unceasing heavy rain he collected more material there than in the whole of his expedition before. He began to plan how he would write up his expedition. There would be one book on the fish of the Amazon and Rio Negro, another on the palms of the Amazon Valley, another on the physical geography of the Amazon. There would also be a travel narrative in the manner of Darwin and Humboldt. He returned to Barra, where he was delighted to find Spruce, awaiting transport to take him up-river. But he also found a letter from the British consul in Para, saying that Edward had contracted a severe case of yellow fever and was unlikely to recover. Alfred had assumed that Edward had left Brazil long before, but there had been delays while repairs were carried out to the ship that was to take him home.

Coincidentally Bates was in Para at the time, where he learnt that the specimens Stevens was receiving in London were selling exceptionally well. As a result Bates changed his mind about leaving Brazil. Wallace decided not to go to Para but to continue with his exploration, this time up the Uaupes river. After some days he was stricken by what proved to be malaria, and had to return to Barra. When he felt better he resumed his journey up the Uaupes, reaching the Colombian border at Juarite, where his crew refused to go any further. Spruce came to his assistance and oversaw his safe passage down-river to Barra. Still weak with fever Wallace packed up his specimens and travelled back to Para where he joined a ship bound for London, with all his specimens and other luggage. It turned out to be a nightmare voyage.

One morning, when they were passing through the Caribbean, the captain of the ship announced that there was a fire on board and they would have to take to the lifeboats. The fire destroyed the ship and all its cargo, including Wallace's specimens, notebooks and journals. After ten days, and with hardly any fresh water left, they were rescued by a merchant vessel bound for London from Havana. The new arrivals overloaded the ship and there was not enough food to go round, but eventually they arrived back in England.

Fortunately the prudent Stevens had insured the collection, but the loss to science was great. Wallace rented a house near the Zoological Gardens, where he installed his mother, sister and brother-in-law. He wrote to Spruce, who was still in Brazil, about his experiences, saying that he planned another expedition, perhaps to the Andes or the Philippines and he wrote a number of scientific articles, based on his few surviving notes and recollections. He also spent a considerable amount of time in the natural history section of the British Museum.

In the space of a year Wallace wrote two books and four important papers. He was struck by the limited range of many species; when very similar conditions obtained in two separate places, a species in one range might be different from the equivalent species in the other. He saw this on the great rivers, where the species on one bank were different from those on the opposite bank when the river was broad; but near the source, where the river was narrow and presented less of a barrier, there would only be one species. He turned this observation around and asked whether closely allied but different species were always separated by some natural barrier, like a broad river. His research confirmed his belief that allied species were related by common descent and were not independently created.

Although Wallace was made welcome by the London scientific community, he wanted to make another major expedition. In his autobiography he said that he would have returned to tropical America but did not want to compete with Bates, who was still out there. He considered visiting central Africa but eventually chose south-east Asia as the best region to continue his work, especially the Malayan archipelago. This time he intended to concentrate on ornithology, about which he was no expert. Only Java had so far been explored by naturalists. After various delays he arrived in Singapore on April 20, 1854 to begin eight years of wandering through the Malay archipelago, accompanied by a sixteen-year-old apprentice collector named Charles Allen. Although Singapore island was being rapidly denuded of trees, the debris was an ideal place to find beetles and by the end of May he had sent Stevens about 1000 specimens, of 700 different species. He was still troubled by recurrent bouts of malaria. After some time in and around Malacca Wallace accepted an offer of hospitality from Sir James Brooke, the wealthy and powerful governor of Sarawak, who promised him every facility to carry out his work. The assistant he had brought from England proved unsatisfactory; in his place Wallace chose a fourteen-year-old Malay boy named Ali, who proved invaluable. For the next seven years he would be Wallace's companion, servant and assistant.

One of Wallace's main objects in coming to Borneo was to observe the orang-utan in its natural habitat, and to obtain good specimens of the varieties and species of both sexes thought to exist. He concluded there was only one species, although the individual variations were large. The immense variability within the same species, the remarkable similarity of its structure and behaviour, and the presence of apparently useless characteristics, such as huge canine teeth in a herbivore, he thought must be due to some natural law as yet unknown. Baby orang-utans were amazingly similar to human babies, suggesting that both might be descended from the same ancestral stock. He conjectured that there must be a natural law to the effect that 'every species has come into existence coincident in both space and time with a pre-existing closely allied species'. During the rainy season, when he was kept indoors, he wrote an article providing evidence for his conjecture both from his own investigations and from those of others, notably Charles Darwin. It was a revolutionary idea which should have stimulated controversy but when it was published there was none at first.

In mid June Wallace visited the lush island of Bali, and then crossed the fifteen-mile strait to the similar island of Lombok. Here he found avifauna of the Australian type, while Bali had avifauna of the Asiatic type. The short deep strait between the two islands was a faunal divide. This was most impressive. He also wanted to find some birds of paradise. No European naturalist had ever seen a specimen in its natural habitat. He was advised that the place to find them was the remote island of Aru. The day after their arrival Ali brought him his first specimen, and subsequently many others. Wallace soon concluded that the fauna of Aru was similar to that of New Guinea, of which the interior was then considered too risky to be visited, but there were important differences. He conjectured that at one time Aru must have been part of New Guinea, which in turn must have been joined to Australia at some time.

When mail arrived he found encouraging letters from Bates and Darwin saying they had been impressed by his article, but Wallace was developing his theories further; he too was planning a book on the origin of species. Among the other islands in the archipelago that Wallace wished to visit was Tenate, midway between Celebes and New Guinea. According to his recollections he was lying in bed there suffering from a rather severe attack of malaria, when in the space of two hours he worked out the whole theory of natural selection. His revelation was the mechanism, survival of the fittest, which came to him in a flash of insight. Over the next two evenings he wrote out the theory in full and sent off copies to Darwin and Lyell. He called his essay *On the Tendency of Varieties to Depart Indefinitely from the Original Type; Instability of Varieties Supposed to Prove the Permanent Distinctness of Species.*

When he could, Wallace continued his journey to New Guinea, arriving on May 25, but his fever returned and he had a serious problem with an ankle injury, which became infected. Moreover the son of the sultan came on a visit to the native chief, who monopolised the food supply and collection of wildlife. Altogether the three-month visit to the mysterious island was a disaster, and he returned to Ternate with relief. In mid September some mail arrived, including a letter from Darwin, from which Wallace learnt that what he had written on Tenate coincided with the same theory that Darwin had been labouring over for almost two decades but not published. What happened in London after that has already been described in the profile of Darwin.

There are some who accuse him, pressured by Lyell and Hooker, of robbing Wallace of his rightful claim to priority. However the two versions of the theory had significant differences. Darwin believed that survival of the fittest operated at the individual level, while Wallace thought it acted at the level of varieties and species. Wallace, however, reacted calmly to the news from England. He planned to elaborate his ideas in a better-documented work to be entitled 'On the law of organic change'. He wrote to thank Hooker and Lyell for the way in which the embarrassing situation had been handled. There was no animosity between Darwin and Wallace at this stage.

For the next six months Wallace, despite his precarious state of health, pressed on with his fieldwork on the island of Batchian. This turned out to be one of his best hunting-grounds: when Ali brought him a bird of paradise new to science, he called it the 'greatest discovery' he had yet made. He wrote a paper developing his idea that the Straits of Malacca formed a faunal boundary, with Asian fauna to the west, Australian fauna to the east. Meanwhile Darwin sent Wallace a copy of *On the Origin of Species* as soon as it came out, explaining that a much fuller treatment was in preparation. Wallace described it as the greatest scientific book since Newton's *Principia*.

Six years of Wallace's life had been spent in the Malay Archipelago and he thought he would need six more to complete his scientific research. He knew his health was at risk and there were letters from home urging him to return before it was too late. The faithful Ali was now quite rich, by Malay standards, and wished to settle down with his wife, and so, in need of an assistant, Wallace now returned to Charles Allen, who had matured into a responsible young man who could be sent off on his own to collect material. After various adventures Wallace collected more specimens of birds of paradise, his main objective, and then prepared to return to London. Altogether he had collected 125 660 specimens consisting of 310 mammals, 100 reptiles, 8050 birds, 7500 land shells, 13 100 butterflies and moths, 83 200 beetles and 13 200 other insects. This time no specimens had been lost. Stevens sold the best of them, making Wallace a wealthy man. When he arrived home he found he was famous. The learned societies elected him to their ranks, regardless of his position on evolutionary theory.

Bates had also returned to England. He had earned much less from his collecting but was respected for his research papers, especially one on mimicry among insects, which provided fresh evidence of what natural selection could achieve. Darwin had taken a paternal interest in his career; writing was an occupation that he did not find easy. Nevertheless his travel book *Naturalist on the Amazons* was judged superior to the similar one written by Wallace, who wrote in a more dramatic style but was often rather sparing of the facts. Wallace was busy sorting out his collections and recovering his health until he was ready to accept Darwin's warm invitation to visit Down House. After this historic meeting they met occasionally in London and corresponded frequently, but Darwin never became a friend of Wallace, rather more of a father-figure. Bates and Wallace had a discussion with Herbert Spencer, who coined the term 'survival of the fittest' and applied it to nearly every sphere of social thought, but Spencer was too much of a philosopher to have much in common with the two biologists.

However, Wallace's intellectual interests extended to some of the controversies of the day. He was impressed by phrenology. He held strong views on the development of the human race. Twelve years of arduous travel in the tropics had not given Wallace a high degree of social polish. He spoke plainly and expressed himself frankly. He was not good at interpersonal relationships, liking to be left alone. 'I seldom have a visitor,' he said, 'but I wish him away in an hour.' His natural reserve, awkwardness and coldness of manner struck people as rudeness. It was impossible for him to make small talk. His unpolished manner particularly upset women, and marriage was not one of his priorities. However the time came when he thought he should marry and start a family, and so he needed to get a job. He applied for one at the Royal Geographical Society, but Bates was also a candidate and he was appointed. Wallace made a proposal of marriage to one young lady which was at first accepted but she changed her mind when the plans for the wedding were at an advanced stage.

He had always been close to his sister Fanny, and she reciprocated. When she took up spiritualism so did he and psychical research became a passion with him, as with other eminent men of that period. He also became interested in mesmerism. Spruce had a friend named Mitten, an amateur botanist, who had a daughter named Annie, who was eighteen when she first met Wallace. After a brief courtship they were married in 1856. She provided the peace, solace and security he needed to complete his work, which was increasingly concerned with the promotion of natural selection. Two years later she gave birth to a son. They spent an enjoyable weekend with the Darwin family at Down House.

In 1868 the Royal Society awarded Wallace one of the Royal Medals, a high distinction, something like a scientific knighthood. Only two weeks earlier his mother had died from senile decay. Like his father Wallace was incompetent in managing his financial affairs and lost through unwise speculation most of what he had earned from sales of the material he had collected. He tried to recoup through his popular book *The Malay Archipelago*, which was and remained a success. But another publication, in which he maintained that natural selection could only produce the animal side of human nature, caused a rift with Darwin. Wallace still lacked a regular job; he applied for the curatorship of a branch of the British Museum in Bethnal Green. Wallace was so confident of being appointed that he moved his family out of central London and thus he lost easy contact with the gentlemen of science who met at the Athenaeum and at meetings of the learned societies.

In 1871, after lengthy negotiations, Wallace acquired a long lease on four acres of land about twenty miles south of London. It was a short walk from a railway station from which he could reach central London in not much over half an hour. He instructed an architect to design an imposing house on the land, where the family could live in style. Halfway through his builder absconded with his advance payments, leaving Wallace with endless bills to pay, but the house, called The Dell, was ready in March 1872 and the family moved in. He still had not heard for certain about his appointment as curator of the new museum but then in August Lyell broke the news that it was to be a museum of art, not natural history, and it had been decided to administer it from the centre. Facing insolvency he sold his treasured collection of exotic birds to the British Museum, and tried to earn money by writing, examining and so on. Friends offered him loans. At this point Wallace's scientific reputation began to decline sharply. He continued to subscribe to theories such as spiritualism and mesmerism, which were not generally accepted or backed by scientific evidence. By associating himself with the promoters of such theories his writings lost credibility. He began work on what proved to be a huge treatise called *The Geographical Distribution of Animals*, dividing the Earth into six geographic zones. When the treatise was published it received high praise.

However, his financial situation was increasingly precarious. He lost a bizarre lawsuit against a flat-earther he had tried to debunk, and was faced with bankruptcy. To avoid having to sell his house he persuaded his publisher to pay him a lump sum in lieu of royalties on future editions of *The Malay Archipelago*. At a séance he had received a message from the spirit world telling him to leave The Dell as soon as possible. He sold it, for a reasonable price, and moved to Dorking, further out of London. More trouble arose when he was elected president of the biology section of the 1876 British Association for the Advancement of Science meeting in Glasgow. He included on the programme lectures on psychical research, mesmerism, etc., complete with demonstrations. This ended in uproar, and Wallace made a lot of enemies as a result.

For the third time in two years Wallace moved house, this time to Croydon, much closer to London. He applied for the post of superintendent of Epping Forest, on the other side of London, but a landscape gardener was appointed instead. He had given up applying for university or college posts, and was coming to the end of his savings. Darwin tried to get him a Civil List pension, but his unorthodox political opinions told against him. Eventually he was granted a pension of £200 a year, putting him in the company of James Joule and Michael Faraday. Next he published *Island Life*, a sequel to *The Geographical Distribution of Animals*, which helped to restore his scientific standing. He built a small house in the village of Godalming, near Guildford, where he had as neighbour a boyhood friend from Neath. Nutwood Cottage, as he called it, remained his home until 1889.

Wallace now began to write articles about social and political questions. He advocated free trade and nationalisation of the ownership of land, for example. In the autumn of 1886 Wallace planned a lecture tour of the United States, hoping to earn money for the benefit of his teenage son and daughter. He also wanted to inspect the American flora and fauna. Initially the plan was to go on to Australia and New Zealand afterwards but he found that at his age crossing the Atlantic was quite enough. On the east coast he met some of the most interesting American intellectuals of his time, and was delighted to find spiritualism well established. He stayed in Washington for some months and then set off for the Midwest in springtime, and eventually the west coast, where he had arranged to meet his brother John and his family in Stockton; they took him to see the scenic Yosemite Valley. He was warmly welcomed by the leading scientists of California. After his year in the United States was over he returned via Canada; he wrote another book, this time a popular sketch of Darwinism. He then became caught up in a crusade against vaccination, which did his reputation no good at all. He received an honorary degree from Oxford, and was offered fellowship of the Royal Society but declined the honour. In 1890 he had moved again, to a house in Dorset. Ten years later he had another house built, which he called Old Orchard, close to where he had been living but overlooking

Poole Harbour. This set him back financially and he was again on the verge of bankruptcy.

Wallace now became interested in astronomy and wrote articles about it for the press. He wrote two more books, *Man's Place in the Universe* and *My Life: A Record of Events and Opinions*, his autobiography. Further major honours came his way: the Copley Medal of the Royal Society and the Order of Merit from the Crown. When he lectured at the Royal Institution the auditorium was packed to the ceiling. As he approached his ninetieth year Wallace was becoming more and more mystical. This mysticism had crept into most of his later writings and finally was expressed as a grand philosophy in *The World of Life*, published in 1910, which went through five editions. Two more books on socialism followed later: *Social Environment and Moral Progress* and *The Revolt of Democracy*. Spiritualism was a major interest of his for nearly sixty years but other causes, such as opposition to vaccination, occupied much of his time and energy. Neither of his children had married, and both had returned to live with their parents. Wallace was feeling his age and was troubled by eczema. His mind remained lucid but he had lost his sense of balance. He died on November 7, 1913. For most of the twentieth century he was a forgotten figure but there has been a reappraisal in recent years

THOMAS HENRY HUXLEY (1825–1895)
Thomas Henry Huxley, known to posterity as T. H., was born on May 4, 1825 above a butcher's shop in Ealing, then a small village twelve miles west of London. His father George was described as an active intelligent man with a quick temper and a reputation for obstinacy. He had been teaching mathematics at Ealing School for eighteen years when Thomas was born. Thomas' mother Rachel was already forty; he was her sixth and youngest surviving child. Thomas had her slender build and black eyes and her lightning intuition. Of his three brothers, James was four years older, and the only one for whom he felt any affection. However Thomas became close to Lizzie, the younger of his two sisters; she helped with the first stages of his education. At the age of eight, in 1833, Huxley started at his father's school, then in terminal decline, but left after two years, the sum total of his formal education. He had a hard start to life.

When his father gave up teaching he moved the family to Coventry, where the Huxley family had lived some time previously, and took over a new savings bank, but this was not a success. Huxley's brother James was

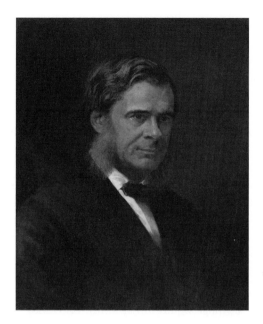

starting a career in medicine and Thomas decided to do likewise. They went to London, where Thomas was apprenticed to a doctor in Rotherhithe, a poor riverside district where fever was endemic. After days spent grinding drugs, he tried to educate himself to the standard required for entrance to University College London. In 1841 he moved in with his sister Lizzie who was living near the college, but changed his plans and enrolled at Sydenham College, a private training establishment for general practitioners named after the surgeon who advised Sloane. There his ability and industry won him a free place at Charing Cross Hospital, London's newest teaching hospital, in the Strand. He threw himself into his work, encouraged by some first-rate teachers. He became a habitué of the Hunterian Museum of the Royal College of Surgeons, where he heard Richard Owen lecture on comparative anatomy.

At twenty Huxley was still too young to obtain a licence to practise medicine, but he needed to earn a living somehow and to repay some of the debts he had accumulated. He decided to join the Royal Navy as an assistant surgeon, and once he was accepted he was selected by the aristocratic Captain Owen Stanley, a lonely, unmarried, grey-haired thirty-five-year-old seafarer who had been to Patagonia, to northern Australia and to New Zealand, amongst other places. Stanley, a good surveyor, was about to be sent on a voyage of exploration around the mysterious island of New

Guinea in Her Majesty's Ship *Rattlesnake*. His mission was to survey the Torres Strait, the passage between northern Australia and New Guinea, and to assess sites for new British colonies.

After a refit the ship was ready to sail before the end of 1846. In the days of sailing ships the prevailing winds were a decisive factor in determining the route to be taken, so that ships bound for Australia normally started by crossing the Atlantic to Brazil before starting the journey east. They ran into severe storms on the way to Madeira where some repairs to the damage were carried out. The next port of call was Rio de Janeiro, where Huxley encountered tropical luxuriance for the first time. From there favourable winds blew the *Rattlesnake* to Cape Town and then on across the Indian ocean to Australia, making landfall at Hobart, the chief town of Tasmania, where the ship underwent some very necessary repairs. After a pleasant week in Hobart they went on to Sydney, where everyone on board had a good time. Huxley enjoyed the social life, and at one of the balls he attended met Henrietta Anne Heathorn, his future wife. Her father was a brewer, based in a village ninety miles south of Sydney. Her mother was the illegitimate daughter of someone in the West Indies by a certain Miss Thomas. She herself was interested in the arts rather than science. During the next three months they became informally engaged to be married once his financial position was more secure.

Stanley purchased a shallow-draft tender, the *Bramble*, for use in surveying close to shore. By this time *Rattlesnake* and *Bramble* had started surveying the coast north of Sydney, as far as Brisbane, returning to Sydney early in the new year of 1848, before setting off again to survey the Great Barrier Reef and to make an expedition into the interior. Huxley accompanied this for some of the way, and had his first real encounter with aborigines. After he left the rest of the party ran into serious difficulties, failed to make rendezvous with *Rattlesnake*, and ended with tragic loss of life. *Rattlesnake* went on sailing round the coast to Western Australia and then returned to Sydney for a refit. Huxley was granted three months' leave. His plan was to marry Henrietta, but then leave her in Australia with her family while he returned to England and established himself in London, subject to whatever the Navy might require of him. However it was decided that they would become formally engaged but not marry until his future was more certain.

On May 8, 1849 *Rattlesnake* finally sailed north, this time destined for New Guinea which the party reached a month later. Although a major trade route passed this way there were no proper charts. Unfortunately the unadventurous Stanley shied away from contact with the natives and refused to allow any exploration of the interior. After six weeks of cruising along the coast of the mysterious island, without once setting foot on it, Stanley decided to return to northern Australia, reaching Cape York at the beginning of October. He seemed to be cracking up under the strain, worried about the reefs, the savages and the value of his work. The *Rattlesnake* went back and forth across the Torres Strait to complete its survey and then went back to Sydney before starting the long voyage home. Prematurely aged, Stanley died at thirty-eight, having spent twenty-two years of his life at sea. The decision was taken to proceed directly to England by the fastest route. On May 2, 1850 they started on a course which took them past New Zealand. Here they landed for a short while, encountered hostile Maoris, and then headed for Cape Horn. They rounded the Horn in severe weather, paused in the Falklands, and then headed for home, arriving in Plymouth on October 23, 1850 and Chatham a few days later.

Once back in London Huxley started trying to improve his financial position. He particularly wanted money for the book he would write about his travels, and he wanted to report the results of his biological research to the learned societies so as to become known in scientific circles. He began to build up a reputation as a lecturer at the Royal Institution and elsewhere. He was elected to the Royal Society and was awarded a Royal Medal and other honours, although still only twenty-eight. He applied for positions as they became vacant, including the natural history chair at the University of Toronto, without success. He became interested in another such position in Sydney, which never materialised. After a series of disappointments he at last landed a post as natural history lecturer at the Royal School of Mines, located in Piccadilly. Shortly afterwards he was offered a better position in Edinburgh but decided to remain in London. Later he was appointed Hunterian Professor at the Royal College of Surgeons and served as Fullerian Professor at the Royal Institution.

Huxley now felt sufficiently confident to ask Henrietta to come to London and marry him, after their eight-year engagement. He rented a narrow terraced house in St John's Wood to be their first home. While she was on the way he had further good news and was able to tell her that he was earning £800 a year and had been able to pay off all his debts. They married on July 21, 1855. Darwin invited Huxley to Down House for the weekend. Hooker and Wollaston were also there. Their host used the occasion to try out his controversial ideas on evolution. Although he had been developing them for many years he was still not ready to go public. Back in London

Henrietta gave birth to a son, named Noel, and babies became an annual event, as they did in so many Victorian households. Not long afterwards Darwin received the famous letter from Wallace that persuaded him it was time to publish. He awaited the reception of his book with trepidation.

There was no shortage of opposition to Darwin's theories, as we have seen. Soon it became clear that, in biology, Huxley would be the one to argue Darwin's case. Huxley was the main supporter of Darwin who did more than anyone else to break down religious and obscurantist opposition to the theory of evolution by natural selection. As Darwin anticipated, the main opposition to his theory came from the ranks of the clergy and the laity. Matters came to a head at the annual meeting of the British Association for the Advancement of Science which in 1860 was held in Oxford. The venue was the new University Museum. One of the speakers was Samuel Wilberforce, the worldly Bishop of Oxford. Before a huge, excited crowd he attacked Darwin's theory. Wilberforce assured the audience that there was nothing in the idea of evolution. He then turned to Huxley and with an insolent smile said 'I beg to know whether it is through your grandfather or your grandmother that you claim descent from a monkey.' No one present could remember his precise words or be sure that Huxley made the oft-quoted remark that 'the Lord has delivered him into my hands'. At any rate, the audience called for Huxley to reply and 'a slight tall figure stern and pale, very quiet and very grave – white with anger,' some said – rose to his feet and said that he had listened with great attention to the Lord Bishop's speech but had been unable to discern either a new fact or a new argument in it –

> except indeed the question raised as to my personal predilection in the matter of ancestry – that it would not have occurred to me to bring forward such a topic as that for discussion myself, but that I was quite ready to meet the Right Reverend prelate even on that ground. If then, said I, the question is put to me would I rather have a miserable ape for a grandfather or a man highly endowed by nature and possessed of great means of influence and yet who employs these faculties and that influence for the mere purpose of introducing ridicule into a grave scientific discussion, I unhesitatingly affirm my preference for the ape.

The next speaker was Joseph Hooker, who said that Wilberforce could not have read Darwin's book or he would not have made such a fundamental misinterpretation of his theory. Undoubtedly it was a famous victory for

Darwin's side, and it is a pity that we do not have a verbatim account of what happened.

Huxley and Owen had been bitter enemies for some time; Huxley never lost an opportunity to discredit Owen. Their differences were political as well as scientific. For example, Huxley campaigned vigorously against Owen's plan of creating a national museum for natural history, arguing that instead of concentrating all the natural history collections in one place they should be broken up and distributed among various metropolitan institutions, for example the botanical material should go to Kew, the mineralogical to the Museum of Practical Geology, attached to the School of Mines, of which Huxley was curator. In science matters reached a climax in the famous controversy of which the Oxford debate was one skirmish. Owen, and many others, believed in evolution for non-humans, but saw man as a separate creation. He claimed there were anatomical features in the human brain which were distinctive, and that these could not be accounted for by evolution. Comparative anatomy was Owen's forte, not evolutionary biology, but even there Huxley managed to trip him up. Owen refused to acknowledge defeat and found himself increasingly isolated.

By this time the Huxley's had three children, two daughters and the son Noel. One day in September Noel became ill with symptoms of scarlet fever; the elder daughter was also affected to some extent. Within two days the boy was dead, of what seems to have been typhoid. Another son was born, named Leonard, but the death of their first born made a lasting impression on the parents, especially Henrietta. The Darwins invited them to spend a fortnight at Down House to recuperate, but Darwin himself was a chronic invalid by this time. Huxley decided to move house to try and break Henrietta's morbid thoughts.

Huxley had been giving popular lectures for some years and he now decided to write a popular book, as distinct from a scientific monograph. This appeared early in 1863 as *Zoological Evidences as to Man's Place in Nature,* and was an instant success. Darwin himself admired the clarity and condensed vigour of his disciple's prose. It was a book whose time had come and although there was opposition it was not abusive. The following year he brought out his *Lectures on Comparative Anatomy,* a different type of book, which also became a classic.

By this time Huxley was approaching forty. He and his wife had six children, four daughters and two sons, and there was to be another daughter. Protégés of Huxley were succeeding the old guard to positions of power and influence. Huxley used his abundant energy to lead the movement to establish Working Men's Colleges, and to spearhead the movement to introduce science teaching in schools, on a par with the teaching of classics. He served on no fewer than ten Royal Commissions and numerous committees dealing with education and science. Huxley was a workaholic and by the end of 1871 it was clear that he was endangering his health. Advised that he needed a complete break, he hastily boarded a ship bound for South Africa. At Gibraltar he transferred to another ship, bound for Alexandria, and then travelled up the Nile as far as Aswan. He resolved to take life more quietly in future.

On his return he put this resolution into practice, with the support of Henrietta and his many friends, who rallied round when he got into financial difficulties. He was now forty-eight, and his children were growing up. He was persuaded to undertake a two-month lecture tour in America, and after many delays he set sail in 1876, reaching New York on August 5 to a tremendous reception. However, the tour, although a great success, was too much for him; and on his return he looked tired and worn out. Seven years later Huxley became President of the Royal Society, but he was not strong enough to make full use of the influence which came with the position and resigned after two-and-a-half years. He refused a peerage on principle but was sworn to the Privy Council, too late for the honour to be more than just a recognition of past services rendered to the State.

To try and improve his health he made an Italian tour with Henrietta but returned exhausted rather than refreshed. He shed as many tasks as he could, although he continued to give lectures on special occasions until deafness forced him to stop. The Huxleys moved from London to salubrious Bournemouth, on the English Channel. By this time his brilliant and gifted daughter Marian had sunk into hysterical imbecility. She was being taken to see the great psychiatrist Charcot in Paris when she died of pneumonia en route. Henrietta never got over her tragic loss. Huxley himself suffered from depressive episodes throughout his life. He died following a heart attack on June 29, 1895, and was buried in the family grave in Finchley. As well as the family his simple funeral was attended by the scientific elite and representatives of the various causes for which he had worked so tirelessly.

Huxley wrote a disappointingly brief autobiography, hardly more than a collection of anecdotes. We can learn more from Beatrice Webb, who recorded this assessment in her diaries:

As a young man, though he had no definite purpose in life, he felt power; was convinced that in his own line he would be a leader . . . I doubt whether science was pre-eminently the bent of his mind. He is truth-loving, his love of truth finding more satisfaction in demolition than in construction. He throws the full weight of thought, feeling, will into anything he takes up. He does not register his thoughts and his feelings; his early life was supremely sad . . . there is a strain of madness in him; melancholy has haunted his whole life 'I have never been satisfied with achievement', he said, and consequently his achievement has fallen far short of his capacity. Huxley is greater as a man than as a scientific thinker.

BEATRIX POTTER (1866–1943)

Natural history was a popular hobby for women of the leisured classes, but those who wished to take it more seriously found obstacles placed in their way. For example they were not allowed to attend lectures open to men. We only hear about them when they became famous for a different reason. The subject of the next profile falls into that category; her reputation as a writer of books for children completely overshadows her attempts to be accepted as a serious naturalist.

Helen Beatrix Potter was born in South Kensington, London on July 28, 1866, the daughter of the barrister Rupert Potter and his wife Helen née Leech. She could see the construction of the new Natural History Museum from her nursery window, if she felt so inclined, but neither of her parents encouraged her to study natural history. As she grew up she grew to love two places in the countryside, where she was dispatched for some fresh air. One was Dalguise House, the family's Scottish summer holiday house in Perthshire, the other was Camfield Place, a country estate of over 300 acres in Hertfordshire, north of London, where her wealthy Potter grandparents lived after her grandfather retired from Parliament. Both sides of the family were nonconformists of the Unitarian persuasion, educated and philanthropic. Her Leech grandparents lived in a house called Gorse Hall, at Stalybridge near Manchester, which she also visited occasionally. In 1872 she acquired a brother, Walter Bertram Potter.

Both their parents were interested in art: Millais was a friend of the family. Beatrix loved to draw and paint, particularly scenes from nature. She read the usual children's classics: Lewis Carroll's two Alice books and Edward Lear's *Book of Nonsense*, and later the novels of Maria Edgeworth and Sir Walter Scott. She learned the three Rs from a tutor, but later her

interests were in natural history, science and art. She also kept pet animals, and she kept a diary, which she wrote in code. Dalguise House was sold and replaced by Wray Castle, on the western shore of Lake Windermere. The castle, in mock gothic style, had a fine view of the lake, but was uncomfortable to live in. This was the beginning of her association with the Lake District, where she spent so much of the later part of her life.

Beatrix Potter, in her twenties, was already an accomplished artist–naturalist. Nature photography was another activity she enjoyed. She had been to art school, where her work was pronounced excellent, and some of her drawings had been published. Her enthusiasm for natural history was nothing unusual among Victorian women of her social class. Collecting, identifying and mounting specimens were popular activities; the more serious naturalists, like herself, also examined them under a microscope. She visited the nearby Natural History Museum to study insects, spiders, butterflies and moths. Flower-painting was also something she was good at, but fungi were her passion and to a lesser degree, fossils. The librarian in the botany department, Ann Lorraine Smith, was interested in mycology, particularly lichens. With her assistance Beatrix, who was one of the first to argue for the symbiotic nature of lichens, studied the mycological literature and was able to correct the staff of the museum on some points of identification.

She also visited Kew, then very much a scientific institution, to which the public were only admitted on certain days and at specific times. Scholars who wished to study there had to apply for permission. The director was the authoritarian William Turner Thistleton-Dyer, son-in-law and successor of the younger Hooker. She was introduced to him by George Massee, first president of the British Mycological Society and now second in command at the Herbarium, and granted the reader's ticket she desired. She would have liked to have attended meetings of the exclusive Linnean Society, but only men were allowed, although women might submit papers. Greatly daring she submitted one 'On the germination of the spores of Agarinaceae'; it was well received but required more work before it would be acceptable for publication; she decided to withdraw it instead. This rebuff marked the end of her scientific efforts, but she continued to draw and paint fungi, with great accuracy, for many more years.

Beatrix had not had much in the way of education but she needed an outlet for her creative energy. Although over thirty by this time she still lived with her parents and went on holiday with them. She was not against marriage; her father knew many intellectual and professional families that might produce eligible bachelors but her mother's social ambitions limited the choice. Beatrix wrote a book and sent it to six publishers who all returned it, some wanted it longer, some shorter, most wanted colour illustrations. She decided to publish it privately, exactly as she wanted it to be. It was called *The Tale of Peter Rabbit*. A commercial publisher then became interested and a contract was signed. Thousands of copies were sold; it was reprinted over and over again.

Meanwhile her brother Bertram had left school and matriculated at Magdalen College, Oxford, where he read classics. After an undistinguished university career he married the daughter of a wine merchant in the Scottish border town of Hawick, without telling his parents, and settled down to a life of farming near the idyllic village of Ancrum. His parents were quite unaware of his marriage, as was his sister at first. He was elected to the Athenaeum, having been first proposed for membership when he was about to go up to Oxford.

Although her next book, *The Tale of Squirrel Nutkin*, is a fable like the first the animals are less anthropomorphic, for example they do not wear clothes. Her publisher was delighted to take this on, and all the other books she produced later. She also began to organise the production of merchandise based on the characters of her books. She took enormous trouble on the text, knowing what children would like best, and the illustrations she made were ideal. Her publisher, Norman Warne, already a firm friend, made a proposal of marriage, which she accepted, but after a short engagement he died unexpectedly.

By this time she was financially independent of her parents and so she could afford to purchase a small farm in the Lake District where she lived for the rest of her life. Hill Top was a working farm of just over thirty-four acres, once part of a much larger property; the farmhouse was late seventeenth century. After some very necessary renovations and improvements she settled in, started farming, about which she knew very little, and continued writing. By the spring of 1909 she had fourteen books published. These were producing substantial royalties and there was additional income from licensing merchandise. She bought another farm, adjoining Hill Top, called Castle Farm.

In June 1912 she received another proposal of marriage, this time from William Heelis, a local solicitor, who had acted as her trusted business and financial adviser. He was nearly five years younger than she was, and came of a respected local professional family. Her parents resisted the marriage, as she expected, but with the support of her brother they finally dropped their opposition. This was the point at which Bertram revealed that he had been happily married himself for the previous eleven years. The marriage took place in London in October 1913. Six months later her father died, and six months after that war was declared.

The newly-weds had settled into Castle Cottage, which they remodelled, keeping Hill Top farmhouse as a studio to which she could retreat for painting and writing. She now had to cope with 160 acres under plough, as well as much pasture, and the farm lads had been called up for military service. Her publishers were in serious financial difficulties; she made every effort to help them to survive, by accepting shares and debentures in place of the back royalties she was owed, and relied more on the sale of merchandise for her income. Her brother died in 1918. The National Trust for Places of Historic Interest or Natural Beauty already had large holdings of mountain land in the Lake District. When many small farms failed after the war the trust was able to buy them up in competition with developers wanting the same land for holiday homes. Aided by her husband Beatrix became a significant landowner herself, eventually owning 4300 acres, and let it be known that this would pass to the trust in due course. She began to write books more for adults, rather than children; these were only moderately successful.

After her mother died Beatrix inherited most of the family wealth, but this did not change her mode of life. Already in 1939 she began to experience serious health problems, necessitating surgery. At the end of 1943 she felt her life was drawing to a close, and on December 13 she died peacefully at Castle Cottage with her husband William beside her. Her ashes were scattered above Hill Top, which became a place of pilgrimage for visitors from all over the world who loved her stories. William died in early August 1945, eighteen months after Beatrix.

8 From Fisher to Hirohito of Japan

SIR RONALD FISHER (1890–1962)

Ronald Aylmer Fisher was born in the East Finchley district of London on February 17, 1890. He and his twin brother, who died in infancy, were the youngest of eight children. His father was a partner in the firm of Robinson and Fisher, auctioneers of fine art, with a reputation comparable to that of Christie's and Sotheby's. Although most of that side of the family were in business, his paternal grandfather was said to have been inclined towards a scientific career. His mother Katie née Heath was the daughter of a successful London solicitor noted for his social qualities. Her only brother also became a lawyer, emigrated to America and became sheriff of Rawlings, Wyoming in the days of the Wild West. One of her sisters married Katie's husband's brother who had been placed high in the Cambridge mathematical Tripos and went into the church. Fisher's highly intelligent but financially naïve father came from a family where intellectual brilliance was often accompanied by problems of personality. The Fishers were brilliant, said one of them, some were dull, some very sane and responsible, some were brilliant but went off the rails, some just went off the rails. Fisher's mother seemed cold and insensitive to outsiders, selfish, indolent and arrogant to relatives. Most likely his personality traits came from her side of the family, but perhaps also from the father's side. Like his mother he suffered extreme myopia, common in people of high general intelligence.

When the boy was six the family moved into Heath House, a mansion designed by his father, set in five acres of parkland and gardens near the top of Hampstead Hill. They also enjoyed the use of a succession of homes in the country. His mother died suddenly of acute peritonitis when he was fourteen, and then his father lost so much money in his business that the mansion was sold and the family moved to a more ordinary home in South London. Fortunately his son was able to win scholarships to help pay for his education. Prodigies generally appear in mathematics, music or linguistics: Fisher was a mathematical child prodigy. At school he had the disconcerting habit of producing the correct answer to a problem without showing how he arrived at it. In later years others found his work difficult to follow, and

criticised him for inadequate proofs and the use of intuition. As with many mathematicians, Fisher's special ability showed at an early age, and love of the subject dominated his professional career. At school he was fortunate in having unusually good teachers. He was physically tough – he was keen on running – but his poor eyesight caused him to be forbidden to work by artificial light; this gave him exceptional ability to solve mathematical problems in his head, also a strong geometric sense.

In 1909 Fisher went up to Cambridge from Harrow School as a scholar of Caius College. He was popular with his fellow undergraduates, and enjoyed himself greatly. Inspired by Galton's Eugenics Society in London he founded one in the university, at a meeting of which he gave a short talk on the subject of Mendelism and biometry, which concluded that

> Biometrics can effect a slow but sure improvement in the mental and physical status of the population; it can ensure a constant supply to meet the growing demand for men of high ability. The work will be slower and less complete than the almost miraculous effects of Mendelian synthesis but on the other hand it can dispense with experimental inbreeding and only requires that the mental powers should be closely examined in a uniform environment, for instance of the elementary schools, and that special facilities should be given to

children of marked ability. Much has been done of late years to enable able children to rise in their social position; still we may as well remember that such work is worse than useless while the birth-rate is lower in the classes to which they rise than in the classes from which they spring.

After three years Fisher graduated in the Wrangler class and stayed on for post-graduate work in physics. At the same time he also developed a strong interest in biology. Advised that he was studying too much he went to Canada to work on a farm near Winnipeg, which was a rewarding experience. He thought seriously about becoming a farmer, but recognised that he did not have sufficient capital to get started. Instead he took a temporary job as statistician at a financial house in the City of London but was soon asked to leave, because his dress was inappropriate. When the First World War began he volunteered for service in the regular army, having been a keen member of the Officers' Training Corps at university, but was shocked to be rejected because of his poor eyesight. During the war Fisher worked in schoolteaching, replacing men called up for military service, but he was a poor teacher and a poor disciplinarian, for he lacked the resonance with students that a teacher needs to get a sympathetic hearing, and he failed to arouse curiosity in his subject. So he battered his head against a brick wall of boyish mischief or against passive incomprehension.

At university he had become friendly with a freethinking young intellectual named Gudruna, daughter of the head of a missionary training college in London. By this time she was married with two children, but her younger sister Eileen, who was only sixteen, was also attracted by Fisher. She was surprised when, quite suddenly, he asked her to marry him. Her father had died and her mother thought her too young for marriage but as soon as she turned seventeen, without her mother's knowledge, she married Fisher from Gudruna's home. Fisher had started teaching at Bradfield College, near Reading. He rented a cottage on the Bradfield estate, with some adjoining land, where he planned to try subsistence farming. Eileen found she was expected to do the routine farm-work, aided by Gudruna and occasionally by Fisher's father. Gudruna brought her three-year-old daughter with her; when it was time for her to start school Gudruna left the cottage and founded a school of her own.

Once the war was over Fisher gave up schoolteaching and started applying for jobs where his professional knowledge would be more relevant.

He tried for a fellowship at Caius, his old college, but without success; in 1920 he was to be awarded a non-resident fellowship at Caius, which would enable him to maintain contact with the university. The previous year he was offered temporary work at Rothamsted, the agricultural research station near Luton. At the same time Karl Pearson, the head of the Galton Laboratory, offered him a position on terms which Fisher considered unacceptable. He accepted the Rothamsted offer, which was soon made permanent, and the following decade was the most fruitful of Fisher's career, when he built up his international reputation in genetics. Two important books date from this period: the revolutionary *Statistical Methods for Research Workers*, first published in 1925, and five years later, *The Genetical Theory of Natural Selection*. This difficult book contains what Fisher called the fundamental theorem of natural selection which states, more or less, that this principle should result in individuals which are well designed to produce as many offspring as possible given all the circumstances. It also contains his thoughts on the future of the human race.

The Fishers needed an adequate and permanent place to live, close to his work at Rothamsted. There were already three children to accommodate; eventually there were to be eight, two sons and six daughters. He bought Milton Lodge, a ten-room Victorian house with a large garden where there was room for everyone and everything. The experimental station was just a fifteen-minute walk across Harpenden Common.

In 1929 Fisher was elected to the Royal Society, in the year he was first nominated, and as a mathematician he was very proud of this, appreciating the influence which comes from being a fellow of that prestigious body. He enjoyed the company of other scientists and was a regular participant at the dining club of the Royal Society, to which he was elected in 1936. In 1933 he was appointed Galton Professor at University College London in succession to Pearson, with whom he maintained a long-standing feud. Pearson's department of applied statistics had been split in half on his retirement, so that Fisher presided over one section and Haldane over the other. The Galton Laboratory offered facilities for the experimental breeding of animals which had not been available at Rothamsted. He installed a mouse colony, which he had started at home, also snails, dogs and even marsupials. He took over the editorship of the *Annals of Genetics* from Pearson, which under his guidance became a journal of importance in statistics. He also published another book, *The Design of Experiments*, which was to become a standard work.

During the Second World War his department was moved out to Rothamsted and the Fishers returned to Milton Lodge. Unfortunately he had for many years been extremely critical of his wife, who was the victim of his volcanic rages, although she appeared only too anxious to support, help and please her husband. To all appearances his accusations were wildly unfair and his anger a cruel and sadistic persecution. The dislocation and its attendant difficulties was the last straw for Eileen and she decided that a permanent separation was the only solution. On top of this the Fishers heard that their eldest son George, a pilot in the Royal Air Force, had died in a plane crash.

In 1943 Fisher returned to Cambridge as Arthur Balfour Professor of Genetics, and built up a flourishing school of mathematical genetics. The professor had an official residence, adjacent to the department, where he lived. He was elected to a professorial fellowship at Caius, his old college, where he became a legendary figure to generations of undergraduates. In 1955 the fellows elected him president, a largely social role which he carried out with great success. Fisher was awarded many high academic and scientific honours and was knighted in 1952. After retirement from his Cambridge chair in 1957 he led a nomadic kind of existence until he settled in Adelaide, the capital of South Australia, where he found the climate and intellectual atmosphere suited him. In his later years he often favoured Indian dress. He had enjoyed excellent health throughout his life until he developed bowel cancer and died in 1962 following surgery.

Fisher was difficult to know, to some natures baffling, to others intolerable. He was at once exceedingly self-centred and utterly self-forgetful. He spoke with clear diction and incisive slow delivery. His conversation was not always easy to follow but always fresh and stimulating. Fisher took immense pleasure in the process of thinking, the play of ideas and the solution of puzzles. In strange company he could be awkward and tongue-tied. Where savoir faire was demanded he might lose his head and have recourse to rudeness. At petty annoyances he was liable to go berserk. He was absent-minded and negligent of routine details.

JOHN BURDON SANDERSON HALDANE (1892–1964)

J. B. S. Haldane and R. A. Fisher had much in common. To start with they were near contemporaries. Both carried out research in genetics. Both worked in the Galton Laboratory at some stage in their careers. Haldane was an able mathematician but Fisher was in a different class altogether. Both were nonconformists but Haldane was a Marxist for much of his life, which

Fisher most certainly was not. They both had a reputation for rudeness, and bad temper; neither was any good at administration.

Haldane was born in Oxford on November 5, 1892. His father was the distinguished Oxford physiologist, John Scott Haldane. Both his parents came from Scottish families noted, for many generations, for energy, enterprise and outstanding intellectual achievement. Their son's intelligence and prodigious memory were recognised early. While still at school he helped his father in research on the physiology of respiration, in particular those aspects of it concerned with deep-sea diving and safety in mines. He was initially very unhappy at Eton, due to his intellectual arrogance, but in spite of this carried away many prizes in the sciences and went up to Oxford with a mathematics scholarship at New College. After reading mathematics in his first year he transferred to 'Greats', which meant classics and philosophy, with the intention of switching back to science later. His interest in these subjects lasted until the end of his life, when he could still quote the Greek and Latin classics from memory. He was fluent in French and German. An early enthusiasm for genetics led to a paper written while he was still a student.

The First World War started two months after Haldane graduated, with a first class degree. He served in the Black Watch, for both sides of his family were Scottish and he had spent much time as a boy at Cloan at the foot of the Ochills. According to his own statement, in an essay called 'Illnesses that make us healthier', he enjoyed the opportunity of killing people and regarded this as a respectable relic of primitive man. He argued that because he unashamedly realised this he was able in later life to let off steam in other ways and was safeguarded from holy anger and pious grief. After being wounded in France in 1915, he returned there briefly before being sent to Mesopotamia. He was again wounded and sent to India to convalesce; after recovery he ran a bombing school there. Haldane had a robust constitution but two war-wounds and other injuries caused him intermittent pain throughout his life.

After the war he returned to Oxford with a New College fellowship, and straightaway published six papers on genetics, presumably written in India. He decided he would teach physiology, although he had no degree or other qualification in the subject. Some of his pupils gained first class degrees and went on to become famous scientists. He was on the fringe of the group of nonconformists who enjoyed the hospitality of Lady Ottoline Morrell at Garsington Manor. Among the friends he made there were the Huxleys, the writer Aldous and the biologist Julian.

Haldane was studying the action of enzymes, which might be regarded as biochemistry, the chemistry of living things. In 1923 he was appointed to the newly created readership in biochemistry at Cambridge, which he was later to combine with a post at the John Innes horticultural research station, and later still with the three-year Fullerian professorship of physiology at the Royal Institution as well. Shortly after he arrived in Cambridge he produced a set of ten papers 'Mathematical contributions to the theory of natural selection' which is considered his best work. He was provided with rooms in Trinity College, although not a fellow, and one day a journalist named Charlotte Franken came to interview him. They took to each other and she decided to divorce her husband and marry Haldane. In those days the legal routine in a divorce case was for the appearance of adultery to be demonstrated, not actual adultery. This was accomplished, with the help of a private detective, but then a university committee, called the Sex Viri, invoked a statute which meant that Haldane was deprived of his post for what they regarded as gross or habitual immorality. Haldane refused to resign, appealed successfully to the courts against the committee's action, was reinstated and the marriage took place. If Haldane had not been in the public eye before he certainly was now.

In 1932, a vintage year, he was elected to the Royal Society. He published no fewer than fifteen scientific papers and two books, *The Inequality of Man*, a collection of his popular writings about science, and *The Causes of Evolution*, where he argued that while Darwin's theory had been modified by subsequent research, the mechanism behind it was now properly understood. He also made a tour of the United States.

The following year Haldane resigned from his Cambridge post in order to take a chair at University College London, where he was to work for almost a quarter of a century. He recruited Boris Chain and others driven out of Germany by the Nazis. Initially Haldane was professor of genetics, but in 1936 he was appointed to the more prestigious Weldon professorship of biometry, the only one of its kind in Britain, and resigned from the staff of the John Innes. One of his research assistants was the biologist Helen Spurway, who said early on that she intended to marry Haldane, and eventually did so. A woman of irrepressible energy, she was able to expound her views at great length, at top pitch and with a ferocity that was not easily quenched.

As Weldon professor he attracted first-class research students, such as John Maynard Smith (1920–2006). Maynard Smith was trained as an engineer but after the end of the Second World War he was attracted to biology and studied evolution and genetics at University College under Haldane, who saw that he had research potential. He obtained an assistant lectureship in zoology, under Peter Medawar, and in 1965 was recruited to lead the school of biological sciences at the new University of Sussex. As he settled into his professorial role Maynard Smith had the circumstances and the confidence to make his own contribution to evolutionary theory, concentrating on evolutionary anomalies, such as the causes of ageing. Evolutionary game theory became his most distinctive and celebrated contribution to evolutionary thought.

Haldane became a Socialist while a student at Oxford; later he described himself as a 'lukewarm supporter of the Labour Party'. By 1937 he was a dedicated Marxist and he joined the Communist Party in 1938, partly because he was impressed by the educational system of the Soviet Union which in the 1930s was one of the leading centres for research in genetics. He regarded Trofim Lysenko as a very fine biologist though some of his ideas were quite wrong. When he left the Party again ten years later it was in protest at Stalin's interference with science, and by 1955 he had turned his back on Communism. He had been out to Spain at the height of the Civil War, where he had some early experience of the effect of air raids. When he returned to England he warned the government, in no uncertain terms, of the necessity of taking air raid precautions seriously, since they would be needed in the forthcoming war. When the war broke out he offered his scientific experience to all three services, and did valuable work for the Royal Navy in particular. His department was evacuated to Rothamsted, as Fisher's had been.

In the spring of 1945 Charlotte had at last divorced him, on grounds of desertion, and he quietly married Helen Spurway. This second marriage was a great success, although neither produced any children. They decided to emigrate. The United States would not admit former members of the Communist Party, but during the First World War he had developed a love for India. In 1957, although he had not yet reached retiring age, he resigned from University College in order to become a member of the biometry research unit at the Indian Statistical Institute in Calcutta, where there were good facilities for research in genetics and biology, not only for himself but also for Helen. As at the John Innes things went smoothly at first but then friction developed due to clashes of personality. He resigned and, with several of his colleagues, set up a research unit in his own house. In 1961 he became an Indian citizen and a year later made his final move to be head of a laboratory of genetics and biometry at Bhubaneshwar, the chief city of the province of Orissa. As an old man he remained an impressive figure, especially in Indian dress, but his health was declining. His death on December 1, 1964 was caused by rectal cancer. There were suggestions that this might have been hastened by his tendency to experiment on himself, but the adverse effects of this seem to have been temporary.

In politics Haldane became well known internationally through his many articles, lectures and broadcasts, which earned him a steady income. At public meetings his ability to speak loudly and simply was invaluable. He wrote fluently and effectively, often on social and political matters, and he also wrote some entertaining verse. Although primarily a research scientist, Haldane regarded the teaching role as just as important and held teaching posts for most of his career. He was at his best with groups of research students. To these he was immensely stimulating because of his wide interests, the range of his knowledge and his wit. When in a cheerful mood, and with people he thought sufficiently well informed to appreciate his somewhat allusive humour, he could be funnier than most professional comedians. Obituaries published soon after his death said that he was habitually rude to unoffending people. He was often abstracted or preoccupied but if approached sensibly gave a sensible response. Those who did not like him were apt to say that he was childlike in his liking for praise and immature in his occasionally violent reactions. Many of the obituaries mentioned these reactions to apparently trivial harassments, usually brought on by someone's foolish behaviour. Haldane received honours from many universities and scientific institutions throughout the world, although not the knighthood which some people think he deserved.

ALFRED KINSEY (1894–1956)

Alfred Charles Kinsey was born on June 23, 1894, in the teeming industrial city of Hoboken, on the Hudson River opposite Manhattan. His harsh and puritanical father Alfred Seguine Kinsey had worked his way up to become instructor in mechanical engineering at the Stevens Institute of Technology in Hoboken. His mother Sarah Ann (née Charles) was the daughter of a carpenter; she was as deeply religious as her husband but poorly educated. Alfred was their first child; later a daughter Mildred Elizabeth and another son Robert Benjamin completed the family.

Kinsey had an unhappy childhood. He suffered from rickets, which left him hump-shouldered for life, and rheumatic fever, which doctors feared had damaged his heart. He also caught typhoid. As the result of these illnesses he often missed school and was unable to take part in the normal rough and tumble of boyhood. His mother was over-protective while his father, a hard man, was slow to praise, quick to find fault. Life began to improve for their son when he was ten years old and the family moved away from unhealthy Hoboken to the suburb of South Orange, set in pleasant countryside west of Newark.

Kinsey had hopes of becoming a professional pianist and was always practising the piano. His style of playing was tempestuous. There was little social life at home and he made no friends. At the local high school his main academic interest was botany, outside school it was birdwatching. He also joined the Boy Scouts, and took up hiking with such determination that his parents sent him to see a doctor, who decided that after all his heart had not been damaged by rheumatic fever. He was more or less a loner, unsure of himself socially. He had an obsession with personal hygiene. Adolescence was the time when he became skilled at concealing his inner life. Later the sickly boy became a big husky fellow, a rugged hardy man looking much younger than his age.

When Kinsey graduated from high school he wanted to go to college and study biology. However his father decided that he should go to the Stevens Institute where he taught and be trained as an engineer. After two years of this Kinsey left Stevens and enrolled at Bowdoin, a small liberal arts college in Maine with a good reputation for biology. He soon earned a reputation as a deadly serious student, determined to succeed. Although he joined a fraternity he was ill at ease in the company of others, unable to make small talk. However he shone in the debating society, was elected as class president, and graduated summa cum laude.

Kinsey left Bowdoin with a Harvard scholarship and began post-graduate work at the celebrated Bussey Institute, where he was determined to make a name for himself. He specialised in the study of gall wasps, from which he expected to learn something new about evolution. When he submitted his Ph.D. thesis on these tiny insects he was awarded a travelling scholarship which he used to make an extended tour of the south-western states, collecting the eggs of different species of gall wasp. Several institutions were interested in recruiting this promising young biologist but in the end he chose Indiana University which admitted students of mixed abilities.

At the university assistant professor Kinsey soon made himself unpopular. He was disrespectful of his elders, prickly, domineering and stubborn. In conversation he could suddenly turn hostile. There was something in his personality that made him want to control other people. He made an effort to build a social life for himself through nature hikes and camping. In this way he met Clara Bracken McMillen, a brilliant undergraduate majoring in chemistry. Friendly, competent and self-possessed she was unafraid to make the first move. They were married in 1921 and went on a camping honeymoon in the White Mountains of New Hampshire. They were to have three children, first a son, who died at an early age, then a daughter Ann, and finally a son Bruce, whose personality traits resembled his father's. Kinsey was very fond of children, especially Bruce.

Kinsey wrote several biological textbooks to supplement his income but also to promote his ideas on evolutionary biology, which came in for criticism. He loved entomological field trips, amassing vast quantities of gall wasps, far beyond any reasonable scientific purpose. He enjoyed summer camps particularly, and on one of them he fell head over heels in love with a young man, the first of several such affairs in the years to come. This seems to have been a turning point in his life, when his obsession with gall wasps became replaced by something quite different.

When the long-serving president of Indiana University retired his successor introduced many changes. One was to offer sex education to the senior students and Kinsey volunteered to take charge of this. He began by teaching a not-for-credit marriage course, in which he took a matter-of-fact biological approach, intended to liberate the young men and women from sexual repression. As part of the course they were asked to complete detailed questionnaires, the precursors of Kinsey's later case histories. Finding these questionnaires inappropriate and open to errors of interpretation he began conducting face-to-face interviews. Soon he was actively engaged in counselling individual students about their sex lives. Kinsey impressed students on the marriage course as earnest and sympathetic, in contrast to the arrogant and gruff image he had routinely projected in biology classes. By 1940 the marriage course had been opened to all students, not only seniors, and it became so popular that enrolment soon reached 400 students per semester. Other universities began to put on similar courses. While sex education was important to Kinsey he was increasingly using it as a springboard for the research that was to dominate the rest of his life. The data he was collecting from students was just a beginning. Kinsey's growing involvement in sex research did not go unnoticed. The president came under strong pressure from the trustees of the university to bring Kinsey to heel. He told Kinsey to choose between giving the marriage course, with modifications, and pursuing his research activities. Kinsey resisted strongly, with support from the students if not from his colleagues, but in the end he had to give way and chose to relinquish the course.

Kinsey realised that in order to obtain meaningful results he needed a more general human sample than the university could provide. He began to travel out of town to conduct interviews with additional subjects. At first these trips were only at weekends but as his interest grew his time away from the campus increased and he began to venture further afield. He took particular interest in the gay communities of Chicago and other Midwestern cities. However he never recognised that a proper statistical sample could not be obtained in this way, and this was to cause him trouble later.

During the 1940s Kinsey embarked on the large-scale study of the sexual habits of men and women that was to make him famous. He increased the number of out-of-town interviews and spent long hours interpreting data and training interviewers. Initially his resources were limited and he used his own money to hire staff and to pay expenses, but before long he received his first outside grant, which helped to reassure the president that he was doing something worthwhile. When a committee was sent to Bloomington to report on his activities they were most concerned about Kinsey's personality; one of them described him as the most intense person he ever knew outside of an institute for psychiatry. Never before had they encountered anyone so sure of himself, so frightfully focused, so suspicious and so contemptuous for the work of others. However he impressed them by showing that he could extract the information he needed not only from students but just as well from prisoners at the local jail, and his activities received a favourable report. Kinsey was then allowed to set up an Institute for Sex Research, associated with the university but not part of it.

In 1943 Kinsey received a $23 000 grant from the National Research Council; over the next decade he would receive hundreds of thousands more from the same source. Since some of the money indirectly came from the Rockefeller Foundation Kinsey claimed that his institute was supported by the foundation. This was to have unfortunate consequences later. The funding enabled Kinsey to hire more assistants; he took great care over their selection and trained them up to be loyal disciples. Soon the quantity of data collected was enough to provide the material for a number of publications. The first one, on the human male, took him two years to write. It included frank descriptions of biological functions and was entirely non-judgemental. Kinsey reported his findings simply and directly, correcting several popular assumptions. He reported that extramarital and premarital sex were more common than was generally believed, that nearly all males, especially teenagers, masturbate, that masturbation did not cause mental illness, and that one in three males experienced at least one homosexual encounter during their lifetimes In an attempt to stress the scientific nature of the work rather than its more sensational aspects he decided to give the book to a well-established firm of medical publishers although more commercial firms were in hot pursuit. Also it was published while the State legislature was in recess to avoid trouble for the university from that quarter. Entitled *The Sexual Behaviour of the Human Male*, the first report was a best-seller. Early polls showed that most Americans accepted its conclusions. However there was criticism of its methodology from scientists, while conservative and religious organisations expressed concern. According to Lionel Trilling, as a biologist Alfred Kinsey could not transcend narrow materialistic thinking; there was an awkwardness about his handling of ideas.

The second report, *The Sexual Behaviour of the Human Female*, was also a best-seller. Some of its more controversial findings concerned the low rate of frigidity, the high rates of premarital and extramarital intercourse, the rapidity of erotic response and a detailed discussion of clitoral versus vaginal orgasm. Critics of the first report returned to the attack, questioning Kinsey's methods and motives. He found himself branded as a subversive and accused of furthering the Communist cause by undermining American morals. At the time the tax-exempt foundations were under suspicion of helping to finance Communism and Socialism in America, and the Rockefeller Foundation was one of those liable to be investigated by a congressional committee. The foundation's continuing indirect support for Kinsey's institute made it seem vulnerable. At the university the president and faculty stoutly defended Kinsey but the foundation succumbed to political pressure and the National Research Council followed suit.

Meanwhile Kinsey's personal behaviour was becoming increasingly strange. In addition to more conventional statistical information he was collecting a huge number of images of sexual activities. Institute staff were coerced into acting out his sexual fantasies in the attic of his house. Altogether he collected 75 000 photographs, carefully classified, and 7000 miscellaneous erotica. After a trip to Peru, to photograph a private collection of erotic ceramics, he was taken ill, and ordered to rest. When his heart started to give him trouble he was hospitalised several times. Meanwhile the institute struggled on, supported mainly by royalties from the Kinsey reports. Another report was planned, this time on sex offenders, so prisoners convicted of sex offences were interviewed, usually by Kinsey himself.

In 1955 he went to Europe for the first time. In London he was horrified by the behaviour of prostitutes and their clientele. Such behaviour, he thought, could only be due to sexual repression; he concluded that a lot of American attitudes must have come from Britain. On the continent he saw that other nations had succeeded in managing human sexuality with less repression, guilt and pain than the United States. After returning to Bloomington he went back to work, against medical advice, saying 'If I can't work I would rather die.' That happened on August 25, 1956, when Kinsey died of pneumonia and heart complications at the age of sixty-two. Science, it was said, was not just his profession, it had become his religion. His two major works broke fresh ground in the field of sex research and led to more open and honest investigations of sexual practices, particularly in the United States. The Institute for Sex Research, which he founded, continues to flourish, under a new name.

DAME HONOR FELL (1900–1986)

One of the most remarkable women physiologists was the cytologist Dame Honor Fell who for over forty years led research in her field at the Strangeways Research Hospital, of which she was director. She was born on May 22, 1900 at Fowthorpe near Filey in Yorkshire, the youngest in the family of seven daughters and two sons of Colonel William Edwin Fell, retired soldier, minor landowner and unsuccessful farmer, and his wife, Alice Pickersgill-Cunliffe, carpenter and amateur architect, who designed the house at Fowthorpe where Honor grew up. Her father knew a lot about horses, from his time in the army, which may have led to her early interest in biology. At that time not many schools for girls included science in their curriculum. One of these was Wychwood School, in Oxford, where Honor was sent to begin her education; and another was Madras College, St Andrew's, where she went next.

From school she went first to Edinburgh University, where she graduated with a B.Sc. in zoology in 1923. She moved to Cambridge in 1923 to become scientific assistant to the physiologist T. S. P. Strangeways, while completing her Ph.D. They worked together on biomedical research at the

Cambridge Research Hospital until Strangeways died suddenly three years later.

The hospital had been founded by members of Strangeways' family, who wanted its work to be continued. They appointed Honor Fell as director of what was then renamed the Strangeways Research Hospital. She held that office until 1970, and during her tenure it developed into a unique institution for research in cell biology. Originally it was housed in a handsome redbrick villa on the southern edge of Cambridge but a grant from the Rockefeller Foundation enabled the laboratories to be extended. Staff numbers rose from thirteen in 1933 to 121 in 1970. Fell always stressed the importance of the application of a range of different disciplines in a research project; in consequence the staff of the laboratory came to include radiobiologists, immunologists, biochemists and electron microscopists. She continually encouraged collaborations between the members of the laboratory and scientists from Cambridge University and other institutions. The hospital attracted visiting scientists from all over the world, who came not only to learn the techniques of tissue and organ culture but also to collaborate with Honor Fell and to gain from her imaginative and enthusiastic approach to biological research. She was deeply interested in the education and training of young scientists, particularly in the third world.

In 1963 she was appointed Royal Society research professor, having been elected to a fellowship in 1952, and became the first woman biologist to be appointed Dame (the female equivalent of knighthood). Among many other honours, medals and prizes she was awarded a D.Sc. by Edinburgh University in 1932, elected to a fellowship at Girton College, Cambridge, in 1955, converted to a life fellowship in 1970, an honorary fellowship of Somerville College, Oxford in 1964, and a life fellowship of King's College London in 1967. She remained a spinster and throughout her working life lived alone, looked after most of the time by her former nanny. After she retired from the directorship she continued to be active in research until the month before her death on April 22, 1986. She stands out as one of the few who managed to surmount the difficulties facing women in all branches of science.

EMPEROR HIROHITO OF JAPAN (1901–1989)

It would be quite wrong to give the impression that biology is a science only studied in the western world, although it is true that the botanists who hunted for plants in remote parts of China and the ornithologists who investigated the rich avifauna of south-east Asia were westerners. On the Pacific shore of Japan, where the warm and powerful Kuroshio current flows from west to east off the Izu peninsula, the fauna and flora of south-east Asia impinge on the north temperate there results an almost overwhelming abundance of species and genera. When Commodore Perry arrived at Shimoda in 1854 with the United States naval fleet he brought two botanists with him, who collected specimens for Asa Gray to identify; he was surprised to find the similarity with the plants of eastern North America which Darwin found so interesting. Since the country opened up to modern science there have been some remarkable Japanese biologists, of whom I have chosen the Showa Emperor Micho Hirohito as one of the best known. Biology, especially zoology, was far more than a pastime to him. His position provided him with resources not available to others while imposing obvious handicaps. I say nothing of his controversial role as ruler of Japan between the wars, and during the Second World War, which has tended to overshadow his scientific merits.

Prince Micho Hirohito was born on April 29, 1901 the eldest son of Crown Prince Yoshihito, who became Emperor Taisho. He was therefore grandson of the Meiji Emperor Mutsohito who brought the Tokugawa

Shogunate to an end and ushered in the modernisation of his country. He was denied much contact with children of his own age. His formal education, according to imperial custom, was in history, languages, literature and art, but one of his teachers, the biologist Hirotaro Hattori (1875–1965), introduced him to the use of the microscope and took him on collecting excursions; he enjoyed the outdoor life. In 1921 he made a visit to western Europe, mainly England, after which he adopted western dress, except on ceremonial occasions. On his return from Europe he married Princess Nagako, became Prince Regent, and at the end of 1926 was proclaimed Emperor.

The previous year a small biological laboratory had been built in the grounds of the Imperial Palace with Hattori as director. The building was enlarged on Prince Hirohito's accession to the throne, after which he usually spent Monday and Thursday afternoons working there and all of Saturday. The Empress usually accompanied him on collecting excursions; she was herself an amateur naturalist who made flower paintings in the classic style. He began by specialising in slime-moulds, later in marine biology, especially the hydrozoa. In the course of the next forty years he made himself an international authority on those lowly coelenterates, requiring enormous patience to understand the microscopic details of their construction. He also carried our the first comprehensive biological survey of the great Sagami Bay, bringing to light 300 new species: who would have guessed that nearly 350 species of crab are to be found there? The flora and fauna of Japan are particularly rich, having suffered little from glaciation.

In collaboration with others Hirohito worked out the floras of northerly Hasu and southerly Suzaki. On state visits to other countries he would usually arrange for visits to botanical gardens and other biological institutions to be included in the programme. He received appropriate honours from learned societies and scientific academies. When his election to the Royal Society was proposed the citation said 'It is by his example as a working scientist and as a person with wide scientific interests that he has exerted great influence and given encouragement to the promotion of science especially in Japan.' He died on January 7, 1989. His elder son, who succeeded him as Emperor, maintains his father's interests in biology, and his younger son is an embryologist and physiologist.

9 From McClintock to Hamilton

BARBARA MCCLINTOCK (1902–1992)

The third daughter of four children, the future Nobel laureate was born on June 16, 1902 in Hartford, Connecticut, and given the first name of Eleanor. Her father Henry McClintock had graduated with a medical degree from Boston University shortly after his marriage, but it took him several years to establish a solid but profitable practice. Her well-bred, adventurous and high-spirited mother Sara née Handy was from an old-established Boston family. After the birth of her fourth child, and first son, she began to show emotional strain. Eleanor developed an adversarial relationship with her mother; to relieve some of the tension she was sent off to live with a maternal aunt and uncle in rural Massachusetts. This was an arrangement that continued on and off throughout the early years of her life; she used to say she was never homesick as long as she was away from her parents. She was happy to roam outdoors, where she developed a love of nature that was to last a lifetime. An inveterate tomboy, she asked to wear boy's clothes at an early age, a wish her parents granted. Once a neighbour who saw her playing sports with boys chided her for this. Sara quickly told the neighbour never to do such a thing again. She began to call herself Barbara, rather than Eleanor, although the change was not formalised until 1943.

At home Barbara McClintock's differences with her mother continued and she grew to be solitary and independent. In 1908 the family had moved to semi-rural Brooklyn, just outside New York City, where her father had obtained a position as company physician to Standard Oil. Interestingly he forbade any of his children's teachers to set them any homework, regarding six hours a day in school as ample time for education, and as a result the future scientist had plenty of time to pursue outside interests, such as playing the piano and ice skating. After graduating from Erasmus Hall High School in Brooklyn, Barbara McClintock enrolled at the College of Agriculture of Cornell University in 1919. At first her mother objected to this, but when her father returned from military service he convinced her not to stand in her daughter's way. During freshman and sophomore years she led the normal college social life of a student, including dating

the boys and even playing tenor banjo in a jazz band. Elected president of the women's freshman class, she was popular among her fellow students but refused an invitation to join a sorority. She never hesitated to flout the social conventions of her time, especially those concerned with woman's role in society. She decided early on to remain an independent single woman devoted to her work; she had little inclination to marry or start a family.

Barbara McClintock became interested in the study of the cell structure, known as cytology, under the tutelage of Lesley Sharp, a professor who gave her lessons on Saturdays. She exhibited a keen intellect as an undergraduate and was invited to take graduate-level genetics courses while still in her junior year. She received her first degree in 1923 and entered graduate school, where she majored in cytology and minored in genetics and zoology. She received her master's degree in 1925 and was appointed instructor after earning a Ph.D. in botany two years later. Then she joined the laboratory of Robert A. Emerson, a pioneer in the genetics of maize, who was drawing promising young geneticists to what became known as the Cornell corn group, among them George Beadle, who went on to win a Nobel prize for his work in molecular genetics. In 1931 Barbara McClintock and her graduate student Harriet Creighton published a landmark study establishing a theory geneticists had previously believed without proof: that a correlation existed between genetic and chromosomal crossovers. Their study revealed that genetic information was exchanged during the early stages of meiosis. These experiments were to become recognised as a cornerstone of modern genetic research.

Despite having achieved worldwide recognition among her peers, the 1930s were difficult years for Barbara McClintock. The Great Depression had forced universities to cut back, and few positions then existed for a woman as a career scientist. Between 1931 and 1933 she worked at various American laboratories with support from a fellowship of the National Research Council. A Guggenheim Fellowship enabled her to visit the Kaiser Wilhelm Institute in Berlin, but she left after a short stay because of her concerns over Nazi racial policies. In 1936 she received her first faculty appointment, as assistant professor of botany at the University of Missouri. At the end of five difficult years, when lack of promotion forced her to look elsewhere, she was invited to spend a year at Cold Spring Harbor Laboratory on Long Island, which was run by the Carnegie Institute of Washington. This was so successful that she remained here for the rest of her career. In 1944 she was elected a member of the National Academy of Sciences, only the third woman to be so honoured, and later that year she became the first woman to be elected president of the Genetics Society of America.

In 1944 Barbara McClintock embarked on the research that was to make her famous. After six years of carefully planned experiments and intensive cytological observations she was able to unravel a complex system of genetic controls that enabled maize to regulate the activity of its genes by transposing portions of its chromosome. When she announced her results in 1950 they were not well received. Compared to molecular genetics her techniques seemed old-fashioned and her results unlikely. Although she continued to publish papers and made several additional attempts to gain the interest of her colleagues at symposia she found few listeners.

Her discovery of transposable genes baffled most of her contemporaries for nearly three decades, but this started to change after 1960. Her pioneering work began to receive due recognition as molecular biologists discovered cases of gene regulation and gene transposition in other organisms. In 1970 she was awarded the National Medal of Science and in 1983 the Nobel prize. Although formally retired in 1967 she continued working with maize until her death on September 2, 1992, shortly after her friends had celebrated her ninetieth birthday. A recluse by nature, she spent nearly fifty years working apart from the mainstream of the scientific community. Although generally a very private person, she could be sociable at times when she loved to talk about science, philosophy and art. She liked to walk local children home from the school bus, describing to them the wonders of nature as they went along.

NIKOLAAS TINBERGEN (1907–1988)

Anyone with the slightest interest in animals knows that some animals instinctively form herds or flocks. Many have a strong territorial instinct. Sexual behaviour and the raising of progeny appears to be largely instinctive. There are all kinds of questions about the nature of this behaviour, which is often specific to the species concerned. Réaumur, as we know, was much interested in the behaviour of insects, particularly bees. In the nineteenth century great French naturalist Jean-Henri Fabre wrote books about the behaviour of beetles, which became very popular. However it is only recently that the study of animal behaviour became a science. The subject of this profile was one of the pioneers in this branch of biology, whose passion was the behaviour of birds

Nikolaas Tinbergen was born in The Hague on April 15, 1907. His parents, whose names were Dirk Cornelius Tinbergen and Jeanette van Eeck, had six children, of whom five survived, four sons and one daughter. They all shared the Protestant work ethic. They were a truly happy family, Nikolaas recalled: 'Our father a liberal-minded very hard-working man with many intellectual and social interests and a fine, if at times somewhat prudish, sense of humour; a man of total honesty; our mother the ever cheerful, understanding, caring centre of "hearth and home".' It seemed like a very typical Dutch bourgeois family, except that two of the sons became Nobel laureates while another who was also exceptionally gifted suffered from depressions and took his own life at the age of thirty-nine. As well as a full-time post at the Gymnasium Tinbergen's father also taught at the business evening school. He was interested in painting and literature. Each summer he rented a country cottage for the family where there was plenty of wildlife to interest his children. Their mother had also been a teacher before marriage; languages were her speciality. Nikolaas took after her in being rather impulsive in behaviour. He grew up with a sense of intellectual inferiority towards his siblings but he was better than them at sport, especially hockey and pole-vaulting. He was also good at drawing and photography. At school his interest in natural history was encouraged

and he joined the Dutch League for Nature Study, which organised field trips for young people. When he left school his parents wanted him to go to university, like all their other children. To help him decide it was arranged that he should spend the summer at Rossitten, a famous bird-ringing station in east Prussia. With his mind made up by this he enrolled as a biology student at the nearby University of Leiden, living at home and commuting by train.

In 1929 Tinbergen was married to Lies Rutten, daughter of a professor of geology at the University of Utrecht. She shared his enthusiasm for natural history. She continued to study chemistry at Leiden until she obtained her degree but did not pursue it any further. Tinbergen, after getting his first degree, decided to stay on for graduate work. His research supervisor advised him against writing a thesis on birds, which might not be thought sufficiently academically respectable, but instead to study a type of bee-hunting wasp found in sand dunes. In 1932 he completed one of the shortest theses on record about this interesting insect and was awarded the Ph.D. at the age of twenty-five. Next he took his young wife on an expedition to eastern Greenland. They settled in an Inuit village, where the people still lived the traditional life of hunter–gatherers. His main professional interest was in the behaviour of the snow bunting, a small bird which arrived to breed in the spring. When they were not around he studied other bird-life and was also very interested in the behaviour of the husky dogs, an essential part of the Inuit economy. He also made a collection of traditional artefacts the Inuit made. Altogether this was a successful year which he described in a book *Eskimoland* he wrote after returning to Holland.

The Tinbergens settled down in Leiden to start a family, where he worked as assistant in the department of geology. He proved himself an excellent and very popular lecturer but his main interest was in field-work, especially animal behaviour. Among the leaders in this field was the German Konrad Lorenz. When they met Lorenz was impressed by Tinbergen's work. They got on well, although their personalities were very different; the German was an egocentric extrovert, the Dutchman modest and retiring. Lorenz invited Tinbergen to spend the spring of 1937 with him at the rather grand country house near Vienna where he lived and worked. This visit was a great success, and they became firm friends. Tinbergen believed it was essential to study the behaviour of animals in their natural habitat, where there were other animals around to influence it, while Lorenz believed in the observation of individuals, some of which he kept at home. Despite this important difference in methodology Tinbergen produced evidence to support some of Lorenz's theories; he was a pragmatist rather than a theoretician himself.

The following year he crossed the Atlantic for the first time, to visit American institutions where research into animal behaviour was being carried out. He found they were strongly influenced by Lorenz's work. Some of the people he encountered had a background in comparative psychology, which gave him some new insights into animal behaviour, but on the whole he was unimpressed by the American work. There were suggestions that he might like to take a position in the United States. This did not appeal to him but at least it persuaded Leiden to promote him to a lectureship.

When the Second World War broke out Holland was not immediately involved. Later, however, the Germans invaded and started to apply Nazi policies. For example, the universities came under pressure to dismiss Jewish professors. There was strong opposition to this, by Tinbergen amongst others, and as a result he was interned in a hostage camp, where the inmates were told they might be shot in retaliation for action taken by the resistance movement. In fact this seldom happened, although the threat was always present. Living conditions were not bad, and the inmates spent much of the time discussing the liberal reforms they hoped to see in Holland after the war was over. Tinbergen was able to maintain contact with German naturalists, including Lorenz, a Nazi supporter, who offered to use his influence to get Tinbergen released. Lies refused the offer, and her husband stayed in the camp for two years.

After the war was over and conditions slowly returned to normal Tinbergen was dismayed to find that Nazi sympathisers were being appointed to high official positions in the post-war administration. He began to think his future might not be in the Netherlands. He already had contacts in Britain, where similar research into animal behaviour was being carried out, and in 1949 he decided to accept the offer of a position in the department of zoology of Oxford University. This was right at the bottom of the academic ladder, much inferior to the post he was giving up in Leiden, but he was assured that he would be promoted as soon as a more senior post became available. There was much ill-feeling at Leiden when he left, and he lost contact with the people he knew there.

After some initial problems the Tinbergens settled down in Oxford and he began to attract research students of high calibre. He was elected to a fellowship at Merton College, but did not seem to make much use of this and after a time he resigned. Within the department he found that there was surprisingly little cooperation between research groups, and perhaps this suited him. He was close to his own group, entertaining them at home and leading them on field trips. A favourite location was Ravenglass Dunes, on the Cumbrian coast, similar to the Dutch coastal region where he had done his first research.

All this time Tinbergen was publishing the results of his observations of the behaviour of birds and his international reputation was growing. His elder brother Jan was also developing a high reputation and it came as no great surprise when he was awarded the Nobel prize in economics. However Nikolaas, a modest man, was not in the least expecting that four years later be would share the prize with Konrad Lorenz and Karl von Fritsch, who had gained remarkable insight into the ways that bees communicate with each other. On the way to Stockholm for the ceremonies in 1973 he paused in Amsterdam for a few days to be awarded the prestigious Swammerdam Medal. Later he was to be awarded other honours. By this time Tinbergen had been promoted to a chair at Oxford and provided with an institute of his own, not too close to the department of zoology where he had worked until then. The college fellowship associated with his professorship was at Wolfson, a new graduate college; again Tinbergen did not make much use of this.

When Lorenz came under attack from certain of the American comparative psychologists, Tinbergen was caught up in the backlash. He resolved to take greater care to distance himself from Lorenz in future. Now that the children were grown up Lies wished him to promote some theories she had formed about childhood autism, although it is doubtful whether she really knew much about the mysterious disorder. In the United States the idea that it was due to a lack of proper bonding between mother and child took a hold. Although entirely without foundation this cranky idea caused much suffering to 'refrigerator mothers', who came to believe that they were at fault. To the dismay of professionals working in the field Tinbergen used his status as a Nobel laureate to promote his wife's theory; they even wrote a book about it, *Autistic Children: New Hope for a Cure*, published in 1983, and he gave embarrassing lectures on the subject, which did nothing for his standing as a scientist.

In 1972 the Tinbergens bought a picturesque cottage near Dufton, in rural Westmorland, which they often used to visit after his retirement, at the statutory age of sixty-seven. This enabled him to make appearances at Ravenglass when research was in progress, and inspire the researchers to ever greater efforts. The depressive episodes that he experienced throughout his life, and which were common in his father's family, became more frequent and more intense. He made a partial recovery from a stroke in 1983, after some weeks in a coma, but was left with memory loss. The end came on December 21, 1988. The Tinbergens had lived an abstemious life and when his will was made public many were surprised to find that he was quite wealthy. Lies survived him by one year.

RACHEL CARSON (1907–1964)

The twentieth century opened up a new avenue of employment for naturalists, particularly in the United States, namely the conservation of wildlife, and this was one which suited women naturalists rather well. Rachel Louise Carson was the best known of these. Through her inspirational writings she helped to change American attitudes towards the world of nature. She was born on May 27, 1907 in Springdale, Pennsylvania, not far from Pittsburgh, just a few weeks after Tinbergen was born in The Hague. Her father Robert Warden Carson was an insurance agent of Scots–Irish stock. Her well-educated mother Maria Frazier MacLean was a retired schoolteacher,

who gave piano lessons to the local children, and stimulated her daughter's love of the natural world and of good literature. Rachel had an elder sister and brother, but since she was much the most gifted of the children, the resources of the family were strained to ensure she had a good education.

By the time she graduated from high school Rachel Carson's ability as a writer and as a naturalist had been recognised. She won scholarships to the Pennsylvania College for Women, an elite liberal arts college in Pittsburgh and was admitted in 1925. In her freshman year her mother came to see her every week, giving her support and encouragement. Rachel Carson was too reserved to participate in organised social functions; sport was her main recreation although she was not especially good at it, and she made some close friends both with her fellow students and with two or three members of the faculty. The most important of these was the head of biology, a dynamic and energetic young zoologist named Mary Scott Skinker, whom she took as her role model. Skinker's career illustrates only too well the difficulties that a woman scientist faced in academia at this time, and so I digress briefly to describe what happened to her. The college president, an able woman, was biased against women in science. Increasingly women's colleges, especially those administered by a female, were recruiting men rather than women to be department heads. Skinker had an M.A. in zoology from Columbia University but needed a Ph.D. She took a year's leave of absence to achieve this at the end of which she found herself unable to advance academically or to support herself in full-time research. Her age (she was in her mid-thirties) was against her, but opportunities had dried up in the depressed economy. She tried government service but discovered that female scientists were not promoted quickly or easily. Completely discouraged, this gifted teacher and able researcher became director of a small private residence for women in New York.

To return to Rachel Carson, in 1929 she graduated magna cum laude in English and biology and was accepted for 'summer research' at the Woods Hole Marine Biological Laboratory on Cape Cod, modelled on the Italian one at Naples. There, seeing the ocean for the first time, she embarked on what proved to be a lifelong study of marine life. After taking advice she enrolled for graduate work at Johns Hopkins University, while teaching there part-time and at the nearby University of Maryland. She arranged that her parents should move to Baltimore; her father's health was declining and in fact he had only a few more years to live. She took her M.A. in marine zoology in 1932 and would have gone on to a Ph.D. but for the need to provide for her mother after the death of her father and care for two young nieces after the death of her eldest sister, whose marriage had not been a success.

She obtained a position at the Baltimore field office of the Bureau of Fisheries, soon to become part of the United States Fish and Wildlife Service, working her way up the Federal bureaucratic ladder more as a science writer than as a field biologist. Successful magazine articles led to her first book, *Under the Sea-Wind*, which came out in 1941, just before the United States entered the Second World War. Because of this the sales were disappointing in spite of excellent reviews. During the war her office was moved to Chicago for two years, then to Washington where the Department of the Interior was based. By 1949 she was publications editor of the Fish and Wildlife Service, with a staff of six assistants, but her official work slowed up her own writing, and she tried without success to find a job as staff writer in the private sector. Her next book, *The Sea Around Us*, published by Oxford University Press in 1951 was an immediate best-seller, as was the reissue of her earlier book. The new book was first serialised in the *New Yorker* magazine; this made her a literary celebrity. She was awarded a one-year Guggenheim fellowship to continue her field research for a seashore guide for the eastern Atlantic Coast.

Royalties from her writings enabled Rachel Carson to resign from government service in 1952 and build herself a cottage on the estuary of the Sheepscot River, near the town of Boothbay Harbor, Maine, where she went to write every summer. This last book in her trilogy, *The Edge of the Sea*, was not quite as successful as the others, although it was rated a best-seller and the *New Yorker* serialised it. She was also broadcasting and writing popular articles. However much of her time and energy was devoted to looking after her aged invalid mother until she died in 1958. By this time she had problems with her own health, a form of arthritis which affected her mobility and, most seriously, malignant breast tumours which were not removed in time.

Rachel Carson then turned her attention to campaigning about pesticide abuse. She came to believe that everything that meant most to her was being threatened by the indiscriminate use of these powerful new chemicals. In her last and most influential book, *Silent Spring* of 1962, she maintained that this threatened humankind and nature with extinction. After a storm of controversy the public responded by calling for the banning of the use of some chemicals and the regulation of others. Americans still struggle to heed her warnings, but her witness for the whole of nature has continued to inspire later generations. She was awarded the Presidential Medal of Freedom posthumously, as well as numerous scientific and literary honours in her lifetime. She died from a coronary on April 14, 1964, just eighteen months after *Silent Spring* was published. Her literary papers were deposited in the library of Yale University.

FRANCIS CRICK (1916–2004)

Francis Crick was born on June 8, 1916 in the Midland town of Northampton, a centre of the shoemaking industry in England. His parents, Harry and Anne Elizabeth Crick née Wilkins, were a middle-class couple who lived near the town, where his father managed the shops in London where the footwear produced in the family's factory was marketed. His mother was the daughter of a self-made businessman, who started a successful chain of clothing stores. The only scientists in the family were his paternal grandfather, who was an amateur naturalist, and a brother of his father's, who was a pharmacist. Francis attended the local grammar school until at the age of fourteen he won a scholarship to Mill Hill, the nonconformist London public school his father and three brothers had attended. At the age of eighteen he entered University College London where he graduated with a second class degree in physics, and then started research under the physicist

E. N. da C. Andrade, supported financially by his pharmacist uncle. He was given a rather dull problem on viscosity to investigate and was working on this when the Second World War broke out.

During the war he was employed as a civilian by the Admiralty, mainly designing measures to protect shipping from acoustic and magnetic mines. In 1940 he married Doreen Dodd, a fellow undergraduate from University College with a degree in English literature, but the marriage did not last long. After they separated Crick shared a flat with his colleague the bohemian eccentric mathematical logician Georg Kreisel, who was born in Austria of Jewish parents. Kreisel had been sent to school in England, before the *Anschluss*, and then to Trinity College, Cambridge.

By the end of the war Doreen and Francis were divorced, with Francis taking responsibility for bringing up their son, Michael, which he handed over to his parents. He had met Odile Speed, who was to become his second wife. She was of French extraction with interests in the visual arts, not science. Although under-qualified to start scientific research in earnest he decided to try, and after considering various possibilities chose molecular biology. He looked for a suitable opening in Cambridge. In 1947 he found one in biophysics at the Strangeways Research Laboratory, directed by Honor Fell. For two years he worked there, supported by a research studentship from the Medical Research Council, after which he was recruited by Max Perutz to join his small research team of molecular biologists, housed in the Cavendish Laboratory.

Crick now felt secure enough financially to propose to Odile. They were married in August 1949 and went to live in Cambridge. A loudmouth with a braying laugh, he annoyed Sir Lawrence Bragg, the head of the Cavendish, who nevertheless thought he had the makings of a good theoretician. He briefly became a fellow of Churchill College but resigned when he found that the college was going to have a chapel; he strongly disapproved of religion in any form. The Cricks moved to 19, Portugal Place, in the heart of Cambridge, where Odile gave birth to a daughter, Gabrielle. The parties they gave there became legendary. Anne Crick, the mother of Francis, who had been looking after Michael, the son by his first marriage, moved from Northampton to a large house in Barton Road, on the western outskirts of Cambridge.

The story of the discovery of the double helix has so often been told that only the barest outline need be given here. It starts with the arrival of the young American James Watson in 1951. Crick and Watson took to each other and were soon collaborating in research. However it was Maurice Wilkins at King's College London, who was generally considered to be leading the investigation of the structure of DNA, using x-ray crystallography. In a few words, the sequence of events was as follows. The famous American biochemist Linus Pauling had written an article about the structure and sent a copy to Crick. Watson took this to London and showed it to Wilkins and to Rosalind Franklin, his assistant. Franklin was very annoyed by Pauling's paper and so Wilkins took Watson into his office where, without Franklin's permission, he showed Watson an x-ray photograph she had taken. Watson, but not Franklin or Wilkins, realised this provided important evidence as to what the structure of DNA must be. He took the news back to Crick in Cambridge and together they worked out the double helical structure, with its ability to replicate. By April 1953 they were ready to announce their great discovery to the scientific world, which opened the way to a much deeper understanding of the fundamental mechanisms which make living organisms possible.

Although generally accepted it was almost ten years before the details were finally confirmed. Crick, Watson and Wilkins were awarded the Nobel prize for medicine and physiology in 1962. Watson's book on the discovery, called *The Double Helix*, seriously underplays Franklin's role, but as Crick said afterwards:

> Rosalind's difficulties and her failures were of her own making. Underneath her brisk manner she was oversensitive and, ironically, too determined to be scientifically sound and avoid short-cuts. She was rather too set on succeeding all by herself and rather too stubborn to accept advice easily from others, when it ran counter to her own ideas. She was proferred help but she would not take it.

Watson thinks she may have been mildly autistic. She died of cancer in April 1958 at the age of thirty-seven.

Crick's finances had been much improved by a series of shrewd investments in Cambridge property. When his mother died in 1955 he replaced Croft Lodge, her large house in Newnham, by a modern apartment building consisting of 20 flats, some of which he sold, others he rented to academic visitors. By 1966 the outlines of molecular biology seemed fairly well established but there was much work to be done in filling in the details. Crick thought it was time to move on and chose embryology, now called developmental biology, as his new research field. He went to the Salk Institute for Biological Studies at La Jolla, California on a sabbatical and made repeated visits in the following years. In 1976 he was operated on for a constricted oesophageal sphincter but made a good recovery and resumed his normal lifestyle. The next year he agreed to join the Salk Institute on a permanent basis and resigned from Medical Research Council employment.

The Cricks enjoyed life in southern California but always spent their summers in Cambridge, where they retained a base. He wrote his own account of the discovery of the double helix, to counter Watson's but his book did not become a best-seller. He had always been interested in trying to understand the workings of the brain in collaboration with the young neuroscientist Christof Koch. In 1991 he was made a member of the Order of Merit. Three years later he was diagnosed with colonic cancer and underwent chemotherapy. He died on July 28, 2004. His scientific papers had already been bought by the Wellcome Trust for a sum that was unprecedented for a living scientist.

WILLIAM HAMILTON (1936–2000)

William Donald Hamilton was born in Cairo on August 1, 1936 to New Zealander parents; his father was an engineer who worked in various countries before retiring to England. Their son was brought up for the most part in a rural and wooded part of Kent, not far from Down House where

Darwin lived the latter part of his life. He described his childhood as idyllic, full of freedom to roam, and of maternal inspiration and encouragement. One day he unearthed some material his father had buried, connected with wartime research on grenades, and severely injured himself when there was an explosion. Surgery saved his life but some of his injuries were permanent. He enjoyed taking risks, especially those involving explosives.

Already he was fascinated by the behaviour of insects. At Tonbridge School he chose Darwin's *On the Origin of Species* as a prize. In 1957 he went up to St John's College, Cambridge, on a state scholarship after a rather dull two years of compulsory national service in the army. The old-fashioned undergraduate course failed to inspire him but the situation changed when he found Fisher's classic work, *The Genetical Theory of Natural Selection*, in the college library. This had not had much impact on the Cambridge biologists, who regarded Fisher as just a statistician. Hamilton was surprised to find that the great man was still in post, awaiting the appointment of his successor as Arthur Balfour professor of genetics. Fisher allowed the third-year undergraduate to work in his department where, as he wrote to his sister Mary 'I begin to think that my ambition to be a theoretical biologist can be more than a dream in spite of my poor mathematical ability.' In fact Hamilton was quite a capable mathematician, although not in the Fisher class, and he was able to grasp Fisher's reasoning, which had a profound influence on his thought.

Already as a graduate student Hamilton had become fascinated by altruism in the social behaviour of animals, which Darwin found so difficult to account for. It was already well established that animal behaviour evolved, just as the organs of animals evolved, but why was altruistic behaviour advantageous to the individual? Hamilton worked out a modification of Fisher's fundamental theorem in which fitness is extended to what he called inclusive fitness. The crucial difference is that it is not only the individual's own offspring that matter, but all the others that the individual helps to come into existence, because it is the genetic material which matters, not the individual. This modified the concept of natural selection so that it satisfactorily accounted for the behaviour of insects like the honeybee. The theory of inclusive fitness, which Hamilton published in 1964, was on the whole well received. Later Hamilton was much honoured, and nowadays textbooks of animal behaviour, behavioural ecology and evolution introduce inclusive fitness as an uncontroversial principle, the only significant extension to Darwinism in the twentieth century.

To begin with, however, Hamilton had trouble finding an institution where he could continue his work. His first regular employment was a lectureship at Imperial College, London, which meant that he was based at the field station of Silwood Park, considered the British entomological Mecca. In 1967 he married Christine Ann Fries, then a trainee dentist, by whom he had three daughters. Ten years later he left Silwood Park, where he was regarded as the resident genius but a poor lecturer, and took his family to the United States. After a period as visiting professor at Harvard he obtained a regular faculty position at the University of Michigan. Six years later the Hamiltons returned to England where he was awarded a Royal Society research professorship, to be held in Oxford. He was elected to a fellowship at New College, where his disciple Richard Dawkins was a tutor, and lived with Christine and their three daughters in the nearby village of Wytham, where the adjacent estate provides the university with a field station for ecological research. However the marriage broke up about 1990 and Christine resumed dental practice in Orkney.

Hamilton met an Italian science journalist named Luisa Bozzi, who became his partner for what proved to be the last six years of his life. He had always been eager to make field trips to exciting locations, particularly in Amazonia. His last such expedition was to the Congo, where he was trying to discover whether the source of HIV could be traced to the wildlife population. He did not find any evidence for this but soon after he returned to Oxford he was taken seriously ill and, after several weeks in intensive care, he died on March 7, 2000. Risk-taking seems to have been part of the Hamilton family ethos and, in the semi-autobiographical collection of his papers entitled *Narrow Roads in Gene Land*, he wrote that he expected to die long before old age, and would not regret it.

Epilogue

Until relatively recently it was not at all easy to earn a living as a biologist. The usual way to begin was by studying medicine, as did Sir Hans Sloane, Carl Linnaeus, John Hunter, Robert Brown, Sir Richard Owen, Charles Darwin, Joseph Hooker, Louis Agassiz, Asa Gray and T. H. Huxley, although except for Sloane and Hunter they practised medicine or surgery only briefly if at all. Some of my subjects earned some money by writing or publishing or illustrating books on natural history, like Maria Sibylla Merian, John James Audubon, John Gould, Alfred Wallace, Beatrix Potter, Francis Galton and Rachel Carson. Some found a benefactor, as did John Ray with Sir Francis Willughby; both Carl Linnaeus and Louis Agassiz were good at finding patrons. Only Sir Hans Sloane, Sir Joseph Banks, Count Buffon, William Hooker, Alexander von Humboldt, Sir Francis Galton and Emperor Hirohito had no real need to earn a living.

The significance of titles differs from one country to another; the United States, of course, has none. In France Lamarck inherited the title of chevalier, while Buffon became a count and Cuvier a baron and then a peer of France as a result of their own efforts. In Britain Sloane and Banks were made baronets, because they served as presidents of the Royal Society. Baronetcies were still being created in Victorian years but were seldom given to scientists: Owen, both the Hookers, Galton and Fisher received knighthoods, and Honor Fell the female equivalent. Another British distinction was membership of the Privy Council, conferred on Banks and Huxley, and of the Order of Merit, conferred on the younger Hooker and Wallace. In Prussia the Humboldts inherited the 'von' distinction, as did Réaumur the 'de' in France, and in Sweden Linnaeus acquired 'von' through his own merits. In most cases such titles are of no great interest, except as indicators of social status.

A better indicator of social origins may be the occupation of the father of the subject, which has more bearing on the family's social status. The fathers of Maria Sibylla Merian, Sir William Hooker, Sir Richard Owen, Audubon, Sir Francis Galton, Sir Ronald Fisher, Rachel Carson and Francis Crick were all in business. The fathers of Charles Darwin, J. B. S. Haldane and Barbara McClintock were physicians. The fathers of Sir Hans Sloane, Ferchault de Réaumur, Count Buffon and Alfred Wallace might be described as lawyers. The fathers of Alexander von Humboldt, Georges Cuvier, Geoffroy de Saint-Hilaire and Honor Fell were retired soldiers.

Carolus Linnaeus, Robert Brown and Louis Agassiz were the sons of clergy-men. Richard Spruce, T. H. Huxley, Alfred Kinsey and Nikolaas Tinbergen were the sons of teachers. The fathers of John Hunter and Gregor Mendel were working farmers. John Ray, Antony van Leeuwenhoek and John Gould were the sons of manual workers. William Hamilton was the son of an engineer. Sir Joseph Banks was the son of a landowner. Only Sir Joseph Hooker was the son of a biologist. In some cases we can be sure, in others suspect, that the mother was an important influence, but informa-tion is generally lacking. Almost all of the subjects married; the exceptions are Alexander von Humboldt, Richard Spruce, Gregor Mendel, Honor Fell, Barbara McClintock and Rachel Carson. Antony van Leeuwenhoek, Jean-Baptiste Lamarck, Louis Agassiz, J. B. S. Haldane and Francis Crick married more than once. Only Gray's and Haldane's marriages were childless.

In Britain, thanks to the influence of men such as Sir Hans Sloane, natural history and especially botany had become highly fashionable by the middle of the eighteenth century, and a polite accomplishment for women of the leisured class, in what Lynn Barber (1980) has aptly called the heyday of natural history. There was a ready market for books about it, both intro-ductory and more advanced, such as Gilbert White's *Natural History of Selborne*. Many of them were written by women, such as Maria Jacson (1755–1839) and Agnes Ibbetson (1757–1823). Ann Shteir (1996) has described the backgrounds of several of these writers, and she also men-tions several women botanical illustrators and collectors, both at home and abroad, of the same period.

On the continent also, women were not welcomed in professional circles, which were exclusively male, with rare exceptions such as the high-born French anatomist Marie Thiroux d'Anconville (1720–1805). Women were denied access to higher education until quite late in the nineteenth century, and after the formal barriers fell, prejudice continued, notably in university faculties and scientific academies. Very few opportunities were available for women to make a career in biology or any other kind of science until the end of the nineteenth century.

The mycologist Annie Lorraine Smith (1854–1937), who specialised in lichens, was librarian at the natural history department of the British Museum and in 1905 was elected one of the first women members of the Linnean Society. Possibly the first woman to obtain an academic position in biology was the palaeobotanist Marie Carmichael Stopes (1880–1958) who achieved double honours in botany and geology at University Col-lege London, and went on to become lecturer on botany at Manchester

University; however, she is better known as a sex educator and advocate of birth control than as a biologist. Since then there have been some remarkable women physiologists, such as the cytologist Dame Honor Fell who, as we have seen, led research in her field at the Strangeways Research Hospital, of which she was director for over forty years. Her near-contemporary the American geneticist Barbara McClintock (1902–1992) was awarded a Nobel prize, while the American ecologist Rachel Carson (1907–1964) became famous for her writings. Hopefully the situation is changing; today women earn more than half the degrees in biological science awarded in Britain, both at first degree and post-graduate levels, and many go on to make successful careers in biological science.

Since several of the profiles feature artists, I conclude this epilogue with some remarks about biological art more generally. Of course we must distinguish between artists for whom natural history was a secondary interest, and artist–naturalists who were highly skilled in making illustrations of plants and animals. The German Joseph Wolf, who worked for John Gould, is considered the greatest of all animal painters. Although fish and insects were not neglected, birds were also a favourite subject. The problem for some illustrators was avoiding a tendency to anthropomorphism. Both Audubon and Edward Lear gave their birds human characteristics, probably unintentionally. Bewick's *British Birds* set a high standard of depicting birds in their natural habitat, as if they were still alive.

Obviously it is much easier to depict a live plant than a live animal, and it is is also easier to obtain a dead plant than a stuffed animal. Although Maria Sibylla Merian was a botanical artist she also depicted caterpillars, with the plants they fed on, and sometimes spiders as well. Flower painting, particularly, attained a high degree of refinement through the work of the Belgian Pierre-Joseph Redouté (1746–1840), whose superb pictures of roses remain much sought after by connoisseurs, yet his work on lilies is considered even more wonderful. However it was the Bauer brothers, Franz and Ferdinand, who brought the art of flower painting to a standard of perfection which has never been surpassed. The senior Bauer, we recall, was the artist selected by Banks for the *Investigator* expedition.

The pictorial representation of animals by artists has a curious history. We have a thirteenth-century drawing of a lion and a porcupine by Villard de Honnecourt, who states that they were drawn from life, but the lion looks more like an illustration from a mediaeval bestiary. Engravings of giant whales washed ashore, said to be drawn accurately from nature, were shown with ears. Plants and animals often appear incidentally in pictures which are primarily concerned with other things but only rarely are they the main subject. When Albrecht Dürer published his famous woodcut of a rhinoceros he had to rely on second-hand evidence, so he used his imagination, and later artists based their representations on his. Although Dürer's study is generally realistic he portrayed a small secondary horn on the shoulders of the animal, and this gave rise to the two-horned rhinoceros, in which the imaginary horn is similar in size to the real one. Human anatomy received particular attention, since it was important for the correct illustration of medical books as well as in the training of artists. Even Leonardo da Vinci made mistakes in his anatomical drawings, for example showing features of the human heart which Galen led him to expect but which are non-existent. Several old masters made another strange mistake in the portrayal of horses, giving them eyelashes on the lower lid, a feature which belongs to the human eye but not that of the horse. A galloping horse was often depicted but the order in which it placed its hooves on the ground was misunderstood until photography showed the true situation. In the eighteenth century George Stubbs (1724–1806) understood the anatomy of the horse to perfection, through years of study of equine corpses in a remote Lincolnshire farmhouse.

Bibliography

PROLOGUE

Barber, Lynn (1980) *The Heyday of Natural History*. London: Jonathan Cape.

Brockway, Lucile H. (1979) *Science and Colonial Expansion: The Rise of the British Royal Botanical Garden*. New York: Academic Press.

Desmond, Adrian (1989) *The Politics of Evolution*. Chicago, IL: University of Chicago Press.

Gourlie, N. (1953) *The Prince of Botanists*. London: H. F. and G. Witherby.

Moore, Wendy (2005) *The Knife Man*. London: Bantam Press.

Nordenskiold, Erik (1929) *The History of Biology*. London: Kegan Paul, French Taubner.

Ridley, Mark (1996) *Evolution*, 2nd edn. Oxford: Blackwell.

Ridley, Matt (1999) *Genome*. London: Fourth Estate.

Segal, Nancy L. (2006) *Indivisible by Two: Lives of Extraordinary Twins*. Cambridge, MA: Harvard University Press.

Shortland, Michael and Yeo, Richard (eds.) (1996) *Telling Lives in Science*. Cambridge: Cambridge University Press.

CHAPTER ONE

John Ray

Raven, C. E. (1942) *John Ray: His Life and Works*. Cambridge: Cambridge University Press.

Maria Sibylla Merian

Davis, Natalie Zemon (1995) *Women on the Margins: Three Seventeenth Century Lives*. Cambridge, MA: Belknap Press.

Wettengl, Kurt (ed.) (1998) *Maria Sibylla Merian: Artist and Naturalist, 1647–1717*. Frankfurt-am-Main: Gert Hatje.

Sir Hans Sloane

Beer, Gavin de (1953) *Sir Hans Sloane and the British Museum*. London: Oxford University Press.

Brown, Martin (1995) *Hans Sloane 1660–1755*. Belfast: Blackstaff Press.

MacGregor, Arthur (1944) *Sir Hans Sloane*. London: British Museum Press.

Antony van Leeuwenhoek

Dobell, C. (1932) *Antony van Leeuwenhoek and his Little Animals*. New York: Russell and Russell.

Schierbeck, A. (1959) *Measuring the Invisible World: The Life and Works of Antoni van Leeuwenhoek FRS*. New York: Abelard-Schuman.

CHAPTER TWO

René-Antoine Ferchault de Réaumur
Torlais, Jean (1936) *Réaumur*. Paris: Desclée de Brouwer.

Georges-Louis Leclerc, comte de Buffon
Fellows, Otis E. and Milliken, Stephen E. (1972) *Buffon*. New York: Twayne Publishers.
Gascar, Pierre (1983) *Buffon*. Paris: Gallimard.
Roger, Jacques (1997) *Buffon*, trans. Sarah Lucille Bonnefoi. Ithaca, NY: Cornell University Press.

Carl Linnaeus
Blunt, Wilfred (1971) *Linnaeus: The Compleat Naturalist*. New York: Viking Press.
Fries, T. M. (1923) *Linnaeus: The Story of his Life*, trans. and abridged by Benjamin D. Jackson. London: H. F. and G. Witherby.
Gourlie, N. (1953) *The Prince of Botanists*. London: H. F. and G. Witherby.
Hagberg, Knut (1952) *Carl Linnaeus*, trans. Alan Blair. London: Jonathan Cape.
Koerner, Lisbet (1999) *Linnaeus: Nature and Nation*. Cambridge, MA: Harvard University Press.

John Hunter
Gray, Ernest (1952) *A Portrait of a Surgeon: A Biography of John Hunter*. London: Robert Hale.
Kobler, John (1960) *The Reluctant Surgeon: The Life of John Hunter*. London: Heinemann.
Ovist, George (1981) *John Hunter 1728–1793*. London: Heinemann.

CHAPTER THREE

Sir Joseph Banks
Cameron, H. C. (1952) *Sir Joseph Banks*. Sydney: Angus and Robertson.
Chambers, Neil (2007) *Joseph Banks and the British Museum: The World of Collecting 1770–1830*. London: Pickering and Chatto.
Gascoigne, John (1994) *Joseph Banks and the English Enlightenment*. Cambridge: Cambridge University Press.
Lyte, Charles (1980) *Sir Joseph Banks: Eighteenth Century Explorer, Botanist and Entrepreneur*. Newton Abbott: David and Charles.
O'Brian, P. (1987) *Joseph Banks: A Life*. London: Collins Harvill.

Jean-Baptiste de Lamarck

Burkhardt, Richard W. (1977) *Lamarck and Evolutionary Biology.* Cambridge, MA: Harvard University Press

Jordonova, Ludmilla (1984) *Lamarck.* Oxford: Oxford University Press.

Packard, Alpheus S. (1901) *Lamarck: The Founder of Evolution.* London: Longmans Green.

Szyfman, Leon (1982) *Jean Baptiste Lamarck et son époque.* Paris: Masson.

Georges Cuvier

Appel, T. (1987) *The Cuvier–Geoffroy Debate: French Biology in the Decades before Darwin.* Oxford: Oxford University Press.

Ardouin, P. (1970) *Georges Cuvier, promoteur de l'idée évolutionniste et créateur de la biologie moderne.* Paris: Expansion scientifique française.

Outram, Dorinda (1984) *Georges Cuvier: Vocation, Science and Authority in Post-revolutionary France.* Manchester: Manchester University Press.

Taquet, Philippe (2006) *Georges Cuvier: naissance d'un génie.* Paris: Odile Jacob.

Alexander von Humboldt

Humboldt, Alexander von (1814–1825) *Personal Narrative of a Journey to the Equinoxial Regions of the New Continent,* trans. with an introduction by Jason Wilson, reprint abridged (1995). New York: Penguin.

Lowenberg, Julius (1873) *Life of Alexander von Humboldt,* trans. Jane and Caroline Lassell. London: Longmans Green.

CHAPTER FOUR

Etienne Geoffroy de Saint-Hilaire

Appel, T. (1987) *The Cuvier–Geoffroy Debate: French Biology in the Decades before Darwin.* Oxford: Oxford University Press.

Robert Brown

Mabberley, D. J. (1985) *Jupiter Botanicus: Robert Brown of the British Museum.* Braunschweig: J. Cramer.

John James Audubon

Adam, Alexander B. (1967) *Audubon.* London: Gollancz.

Sir William Hooker

Allan, M. (1967) *The Hookers of Kew 1785–1911.* London: Michael Joseph.

CHAPTER FIVE

John Gould

Tree, Isabella (2003) *The Bird Man: The Extraordinary Story of John Gould.* London: Ebury Press.

Sir Richard Owen

Rupke, Nicolas (1994) *Richard Owen: Victorian Naturalist.* New Haven, CT: Yale University Press.

Louis Agassiz

Lurie, Edward (1960) *Louis Agassiz: A Life in Science.* Chicago, IL: University of Chicago Press.

Marcou, Jules (1972) *Life, Letters and Works of Louis Agassiz.* New York: Macmillan.

Charles Darwin

Barlow, N. (1958) *The Autobiography of Charles Darwin 1809–1882.* London: Collins.

Beer, Gavin de (ed.) (1974) *Autobiographies by Charles Darwin and Thomas Henry Huxley.* London: Oxford University Press.

Bowler, Peter J. (1990) *Charles Darwin: The Man and his Influence.* Oxford: Basil Blackwell.

Browne, Janet (1995) *Charles Darwin: Voyaging.* London: Jonathan Cape.

Browne, Janet (2002) *Charles Darwin: The Power of Place.* London: Jonathan Cape.

Desmond, Adrian and Moore, James (1992) *Darwin.* London: Penguin.

Eldredge, N. (1995) *Reinventing Darwin: The Great Debate at the High Table of Evolutionary Theory.* New York: John Wiley and Sons.

CHAPTER SIX

Asa Gray

Dupree, A. Hunter (1959) *Asa Gray, 1810–1888.* Cambridge, MA: Belknap Press.

Sir Joseph Hooker

Allan, M. (1967) *The Hookers of Kew 1785–1911.* London: Michael Joseph.

Turrill, W. B. (1963) *Joseph Dalton Hooker: Botanist, Explorer and Administrator.* London: Nelson.

Richard Spruce

Seaward, M. R. D. and Fitzgerald, S. M. D. (1996) *Richard Spruce (1817–1893): Botanist and Explorer.* Kew: The Royal Botanic Gardens.

Sir Francis Galton

Bulmer, Michael (2003) *Francis Galton: Pioneer of Heredity and Biometry.* Baltimore, MD: Johns Hopkins University Press.

Cowan, Ruth Schwartz (1969) *Sir Francis Galton and the Study of Heredity in the Nineteenth Century.* Ann Arbor, MI: University of Michigan Press.

Forrest, D. W. (1974) *Francis Galton: The Life and Work of a Victorian Genius.* New York: Taplinger.

Galton, Francis (1908) *Memories of my Life.* London: Methuen.

Gillham, Nicholas Wright (2001) *A Life of Sir Francis Galton.* Oxford: Oxford University Press.

Keynes, Milo (ed.) (1993) *Sir Francis Galton FRS.* Basingstoke: Macmillan.

Pearson, Karl (I 1914, II 1924, III 1930) *The Life, Letters and Labours of Francis Galton.* Cambridge: Cambridge University Press.

CHAPTER SEVEN

Gregor Mendel

Iltis, Hugo (1932) *Life of Mendel,* trans. Eden and Cedar Paul. London: George Allen and Unwin.

Orel, Vitezslav (1996) *Gregor Mendel: The First Geneticist,* trans. Stephen Finn. Oxford: Oxford University Press.

Alfred Wallace

Brooks, J. L. (1973) *Just before* The Origin: *Alfred Russel Wallace's Theory of Evolution.* New York: Columbia University Press.

Fishman, M. (1981) *Alfred Russel Wallace.* Boston, MA: Twayne Publishers.

Mackinney, H. Lewis (1972) *Wallace and Natural Selection.* New Haven, CT: Yale University Press.

Raby, Peter (2001) *Alfred Russel Wallace: A Life.* London: Chatto and Windus.

Schermer, Michael (2002) *In Darwin's Shadow: The Life and Science of Alfred Russel Wallace.* Baltimore, MD: Johns Hopkins University Press.

Slotten, Ross A. (2004) *The Heretic in Darwin's Court: The Life of Alfred Russel Wallace.* New York: Columbia University Press.

Thomas Henry Huxley

Beer, Gavin de (ed.) (1974) *Autobiographies by Charles Darwin and Thomas Henry Huxley.* London: Oxford University Press.

Bibby, Cyril (1960) *T. H. Huxley: Scientist, Humanist and Educator.* New York: Horizon Press.

Desmond, Adrian (1994) *Huxley: From Devil's Disciple to Evolution's High Priest.* Reading, MA: Addison Wesley.

Beatrix Potter
Lear, Linda (2007) *Beatrix Potter: A Life in Nature.* London: Allen Lane.

CHAPTER EIGHT

Sir Ronald Fisher
Box, Joan Fisher (1978) *R. A. Fisher: The Life of a Scientist.* Chichester: John Wiley and Sons.
Fisher, R. A (1930) *The Genetical Theory of Natural Selection.* Oxford: Clarendon Press.
Yates, F. and Mather, K. (1963) Ronald Aylmer Fisher. *Biogr. Mems. Fell. R. Soc. London* **9**, 91–120.

John Burdon Sanderson Haldane
Clark, Ronald William (1968) *JBS: the Life and Work of J. B. S. Haldane.* Oxford: Oxford University Press.
Dronamraju, Krishna R. (1986) *Haldane.* Aberdeen: Aberdeen University Press.
Maynard Smith, J. (1978) *The Evolution of Sex.* Cambridge: Cambridge University Press.

Alfred Kinsey
Gathorne Hardy, Jonathan (2005) *Alfred C. Kinsey: A Biography.* London: Pimlico.

Dame Honor Fell
Mason, Joan (1966) Honor Fell, in *Twelve Portraits of Cambridge Women*, eds. E. Shils and C. Blacker. Cambridge: Cambridge University Press.
Vaughan, Dame Janet (1987) Honor Bridget Fell. *Biogr. Mems. Fell. R. Soc. London* **33**, 235–260.

Emperor Hirohito of Japan
Corner, E. J. S. (1990) His Majesty Emperor Hirohito of Japan, K.G. *Biogr. Mems. Fell. R. Soc. London* **36**, 241–272.

CHAPTER NINE

Barbara McClintock
Comfort, Nathaniel C. (2001) *The Tangled Field: Barbara McClintock's Search for the Patterns of Genetic Control.* Cambridge, MA: Harvard University Press.

Keller, Evelyn Fox (1983) *A Feeling for the Organism: The Life and Work of Barbara McClintock*. New York: W. H. Freeman.

Nikolaas Tinbergen

Kruuk, Hans (2003) *Niko's Nature: The Life of Niko Tinbergen and his Science of Animal Behaviour*. Oxford: Oxford University Press.

Rachel Carson

Brooks, Paul (1973) *The House of Life: Rachel Carson at Work*. London: Allen and Unwin.

Lear, Linda (1997) *Rachel Carson: Witness for Nature*. London: Allen Lane.

Francis Crick

Crick, Francis (1990) *What Mad Pursuit?* Harmondsworth: Penguin.

Ridley, Matt (2006) *Francis Crick: Discoverer of the Genetic Code*. London: Harper Press.

William Hamilton

Grafen, Alan (2004) William Donald Hamilton. *Biogr. Mems. Fell. R. Soc. London* **50**, 109–132.

Hamilton, W. D. (I 1996, II 2001, III 2006) *Narrow Roads of Gene Land*. Oxford: Oxford University Press.

Epilogue

Crowther, J. G. (1960) *Founders of British Science*. London: Cresset Press.

Davis, Natalie Zemon (1995) *Women on the Margins: Three Seventeenth Century Lives*. Cambridge, MA: Belknap Press.

Judson, Horace Freeland (1979) *The Eighth Day of Creation: Makers of the Revolution in Biology*. New York: Simon and Schuster.

Mayr, Ernst (1982) *The Growth of Biological Thought*. Cambridge, MA: Harvard University Press.

Pinault, Madeleine (1991) *The Painter as Naturalist: From Dürer to Redouté*, trans. Philip Sturgess. Paris: Flammarion.

Shteir, Ann G. (1996) *Cultivating Women, Cultivating Science: Flora's Daughters and Botany in England 1760–1800*. Baltimore, MD: Johns Hopkins University Press.

Image Credits

Nikolaas Tinbergen: courtesy of Lary Shaffer

Rachel Carson: from Lear, L. (1997) *Rachel Carson: Witness for Nature*. London: Allen Lane

Francis Crick: from Ridley, M. (2006) *Francis Crick: Discoverer of the Genetic Code*. London: Harper Press

William Hamilton: courtesy of James King-Holmes/Science Photo Library